YOU CAN BET ON IT!

* What is the average
Scholastic Aptitude Te
The nation's average fo
verbal section and 476
has the highest average
and South Carolina has the lowest with 838.

* What are the odds that your children are watching too much TV?
Children ages 2 to 5 watch about 25 hours of television weekly; ages 6 to 11 watch 22 hours; and ages 12 to 17 watch about 23 hours per week.

* What are the odds that your spouse is having an affair?
An estimated 1.5 percent of married people have had an affair in the past year. Since marriage, 1 out of 3 women and 2 out of 3 men have been unfaithful.

* What are the odds that you'll get the same flight number as a plane that crashed?
Zero. U.S. airlines eliminate the flight number after a crash.

THE BOOK OF

ODDS

Winning the Lottery,
Finding True Love,
Losing Your Teeth,
and
Other Chances in
Day-to-Day Life

Michael D. Shook
and Robert L. Shook

A SIGNET BOOK

SIGNET
Published by the Penguin Group
Penguin Books USA Inc., 375 Hudson Street,
New York, New York 10014, U.S.A.
Penguin Books Ltd, 27 Wrights Lane, London W8 5TZ, England
Penguin Books Australia Ltd, Ringwood, Victoria, Australia
Penguin Books Canada Ltd, 10 Alcorn Avenue,
Toronto, Ontario, Canada M4V 3B2
Penguin Books (N.Z.) Ltd, 182–190 Wairau Road, Auckland 10, New Zealand

Penguin Books Ltd, Registered Offices: Harmondsworth, Middlesex, England

Published by Signet, an imprint of Dutton Signet, a division of
Penguin Books USA Inc. Previously published in a Plume edition.

First Signet Printing, October, 1993
10 9 8 7 6 5 4 3 2 1

This book is for Bobbie,
our beloved mother and wife.
MDS
RLS

Contents

Acknowledgments

We are grateful to the wonderful staff at the Bexley Library who went above and beyond the call of duty in assisting us with our research during the writing of this book. Special thanks goes to Robert Stafford, executive director, as well as Marilyn Battin, Joline Boettcher, Jan Chadwick, Wavelin Pritchard, and Karen Stover.

Our appreciation is extended to the following supporters and contributors: Erin Connor, David Goldman, Kara Gutterman, Sarah Hamilton, Virginia MacGregor, Jeff Meyer, Jon Meyer, Matt Meyer, Connie Mitchell, Mark Mitchell, Nancy Mitchell, Sandy Mitchell, John Powers, Ann Roskem, Will Roth, Carrie Shook, Elinor Shook, Herbert Shook, RJ Shook, Rod Simpson, Ann Bell Taylor, Danny Thomases, Jenny Traeger, and Ed Ziv. Mary Liff did a great job both in typing and organizing this book. Jeff Herman, our agent, did his job brilliantly, and so did our editor, Matt Sartwell, and his assistant, Peter Borland. Together, we had a delightful time writing this book; it's a feeling we hope comes through to the reader.

MDS and RLS

Introduction

You don't have to be a bookmaker to be interested in what the odds are. There are many odds that are helpful to know to enable you to make everyday decisions. For instance, the odds are 9 to 1 that your pizza delivered by Domino's will arrive in less than 30 minutes. By the way, 1 out of 2 orders pepperoni on their pizzas—3 percent of Domino's customers order a pizza with everything on it. If you have a fear of flying, stop worrying. The odds in 1988 were 1 in 2.2 million of being killed in an airplane crash. And if you're worried about being killed in an automobile accident, then slow down. The odds of being killed at 65 mph vs. 55 mph increase by an estimated 31 percent. And once you get there, be careful where you park—34 percent of American car thefts occur in parking lots.

Perhaps you're thinking about applying to the Naval Academy. Well, good luck. Only 1 out of 10 applicants was accepted in 1989. And of the 6,900 applications received at the Harvard Business School, 924 were accepted and 800 applicants enrolled in 1988. Speaking of education, it's expensive. Only 6 percent of U.S. families can afford to send a child to a 4-year private college without financial assistance.

Moving on to sex, did you know that 60 to 80 percent of women who use no method of birth control will have a pregnancy within 1 year? Well then, how safe is a condom? Nearly 14 percent of American couples using condoms are having accidental pregnancies. Only 1 out of 3 sexually active teens uses birth control regularly. With the married folks, the "very happily" married have sex 75 times a year; not-too-happy couples only about 43 times per year. And how loyal are American men and women to each other? Since marriage, about 1 out of 3 women has been unfaithful and 2 out of 3 men have had extramarital sex. For those men who are worried about impotency, keep in mind that the odds are 124 to 1 against it if you're under age 30. By age 40 the odds drop to 51.6 to 1, and by age 50, 12.9 to 1. But by age 75, the odds are 1.2 to 1 that you will be impotent.

Here are some interesting facts about Americans' health: 1 out of 100 Americans is color-blind. 1 out of 50 Americans has an ulcer. 1 out of 25 Americans has asthma. 2 out of 25 Americans have a heart condition. 44 percent of Americans wear eyeglasses. The chances that you will suffer from arthritis are 1 in 7. And by age 65, the percent of Americans who have lost all their teeth is 42, and the average American over age 65 has lost 10 teeth. Nearly 3 out of 4 of the people who own home exercise equipment don't use it as much as they intended.

Is your appetite whetted for reading more? If not, the odds are that you're not a normal person (we have no actual source for making this statement—we base it on a strong hunch). All of the other information contained in *The Book of Odds* includes a source from which our statistical data has been collected. So you may read this book for the sheer fun of it, or you may want to use it as a valuable reference. Whatever your preference, we urge you to read it. We'd like to think we didn't collect all of this valuable information for nothing.

Michael D. Shook & Robert L. Shook

1

PEOPLE
PATTERNS

Home on the Range

1 out of 50 Americans lives on a farm.

Source: *Statistical Abstract of the United States 1987*. Bureau of the Census. U.S. Department of Commerce.

No Sweets

1 out of 33 Americans never eats candy.

Source: News. National Confectioners Association.

Single Seniors

1 out of 25 Americans aged 65 or older is divorced.

Source: "A Profile of Older Americans: 1987." American Association of Retired Persons.

Big Mac Attack

1 out of 20 Americans goes to McDonald's every day.

Source: "Fascinating McFacts." McDonald's Corporation: 1984.

Welcome to America

6 out of 100 Americans are foreign born.

Source: *1980 Census of Population: General Social and Economic Characteristics, U.S. Summary*. Bureau of the Census. U.S. Department of Commerce.

Walk to Work

6 out of 100 Americans walk to work.

Source: Ibid.

Big Boozers

7 out of 100 Americans drink half of the alcohol consumed in America.

Source: Update. Alcohol, Drug Abuse and Mental Health Administration. U.S. Department of Health and Human Services. Fall 1987.

Free Food

2 out of 25 Americans receive food stamps.

Source: *Statistical Abstract of United States 1987*.

Single Adults

2 out of 25 Americans 18 years or older are divorced. Same is true for widowed.

Source: Ibid.

Readers of the Good Book

1 out of 10 Americans reads the Bible every day.

Source: George Gallup Jr. *The Gallup Poll—Public Opinion 1986*. Scholarly Resources, Inc.

No Comprendo

11 out of 100 Americans speak a language other than English at home.

Source: Lewis Lapham, Michael Pollan, and Eric Etheridge. *The Harper's Index*. Henry Holt, 1987.

Cats and Dogs

3 out of 25 households have both a cat and a dog.

Source: Louis Harris. *Inside America*. Vintage, 1987.

Chinese Food

12 percent of Americans choose Chinese food when they dine out.

Source: Ibid.

Unwed Mothers

21 out of 100 American babies are born to unmarried mothers.

Source: *Health United States 1986*. Public Health Service. U.S. Department of Health and Human Services.

Join the Union

22 out of 100 American workers are union members.

Source: *Statistical Abstract of the United States 1987*.

Moving Up

29 percent of American workers want their boss's job.

Source: Lewis Lapham, Michael Pollan, and Eric Etheridge. *The Harper's Index*. Henry Holt, 1987.

Wacky Tobaccy

1 out of 3 Americans has tried marijuana at least once in their lives.

Source: "NIDA Capsules." National Institute on Drug Abuse. November 1986.

Abstainers

35 out of 100 Americans abstain from the use of alcohol.

Source: *Health United States 1986.*

Dining Out

35 out of 100 Americans go out for dinner once a week.

Source: "America in the Eighties." R. H. Bruskin Associates Market Research, 1985.

Wishful Weight Watchers

35 out of 100 American women would like to lose at least 25 lbs.

Source: Louis Harris. *Inside America.* Vintage, 1987.

Homebodies

39 out of 100 Americans never go out to the movies.

Source: Louis Lapham, Michael Pollan, and Eric Etheridge. *The Harper's Index.* Henry Holt, 1987.

Home Delivery

2 out of 5 Americans had pizza delivered to them in the last three months.

Source: Ibid.

Churchgoers

2 out of 5 Americans have attended church or synagogue in the last week.

Source: George Gallup Jr. *The Gallup Poll—Public Opinion 1986.* Scholarly Resources, Inc.

Starting Off the Day

41 out of 100 Americans have cereal for breakfast every day.

Source: *Statistical Abstract of the United States 1987*.

Lonesome Seniors

41 out of 100 American women 65 years old and older live alone.

Source: Ibid.

Catching Up on the News

42 out of 100 Americans watch the late news on television.
61 out of 100 Americans read the daily newspaper.
66 out of 100 Americans read the Sunday newspaper.

Source: Ibid.

Days of Our Lives

43 out of 100 Americans watch daytime television almost every day.

Source: "America in the Eighties." R. H. Bruskin Associates Market Research: 1985.

Don't Stay Home Without One

11 out of 25 American households have a gun.

Source: Louis Harris. *Inside America*. Vintage, 1987.

Aliens in America

1 out of 2 American residents who are foreign born is not a citizen.

Source: *Statistical Abstract of the United States 1987*.

No Bars

51 percent of Americans never go to a bar or tavern.

Source: General Social Surveys, 1972–1987. Conducted for National Data Program for the Social Sciences at the National Opinion Research Center, University of Chicago.

No-Brand Buyers

51 percent of Americans buy generic products.

Source: Louis Harris. *Inside America*. Vintage, 1987.

Coast Dwellers

52 percent of Americans live within 50 miles of the coastal shoreline.

Source: *Statistical Abstract of the United States 1987*.

Two-Car-Plus Households

52 percent of American households have two or more motor vehicles.

Source: Ibid.

Where's Mama? Where's Papa?

53 percent of the black children in America live with one parent.

Source: *The New York Times*, January 28, 1988.

Take a Walk

53 percent of Americans walk for pleasure.

Source: *Statistical Abstract of the United States 1987*.

Household Pets

57 percent of Americans have a pet.

Source: Louis Harris. *Inside America*. Vintage, 1987.

Daily Prayers

57 percent of Americans pray at least once a day.

Source: General Social Surveys, 1972–1987.

Drugs in High Schools

58 percent of American high-school seniors have used illicit drugs.

Source: Update. Alcohol, Drug Abuse and Mental Health Administration.

Bang, Bang, I'm Dead

59 percent of American suicides shoot themselves.

Source: *Statistical Abstract of the United States 1987.*

Cutting Coupons

61 percent of American shoppers almost always look over coupons before going shopping.

Source: *Progressive Grocer*, April 1987.

Vows Taken

63 percent of Americans 18 years and older are married.

Source: *Statistical Abstract of the United States 1987.*

State Grown

64 percent of all Americans live in the state where they were born.

Source: Ibid.

The More You Make the More You Cheat

70 percent of American men who make $70,000 or more a year cheat on their wives.

Source: Lewis Lapham, Michael Pollan, and Eric Etheridge. *The Harper's Index*. Henry Holt, 1987.

Kicking the Habit

71 percent of American smokers have tried to quit.

Source: General Social Surveys, 1972–1987.

The Famous Ronald McDonald

96 percent of American schoolchildren can identify Ronald McDonald. (Who is second to Santa Claus.)

Source: "Fascinating McFacts." McDonald's Corporation: 1984.

Church Members

71 percent of Americans are members of a church or synagogue.

Source: *Statistical Abstract of the United States 1987*.

Avoiding AIDS!

72 percent of Americans prefer not to work around people with AIDS.

Source: *Public Opinion*, July/August 1987.

Unemployed and Up the Creek

3 out of 4 Americans who are unemployed receive no unemployment benefits.

Source: Lewis Lapham, Michael Pollan, and Eric Etheridge. *The Harper's Index*. Henry Holt, 1987.

If You Haven't Tried It . . . Don't

76 percent of Americans have tried cigarettes.

Source: "NIDA Capsules."

Male Suicides

77 percent of American suicides are male.

Source: *Statistical Abstract of the United States 1987*.

Life in the City

77 percent of Americans live in metropolitan areas.

Source: Ibid.

Hello, Daddy!

79 percent of American fathers are in the delivery room when their children are born.

Source: Lewis Lapham, Michael Pollan, and Eric Etheridge. *The Harper's Index.* Henry Holt, 1987.

Would You Take This Man/Woman . . . Again?

85 percent of Americans say they would marry their spouse if they had to do it all over again.

Source: "America in the Eighties." R. H. Bruskin Associates Market Research, 1985.

Male Doctors

85 percent of American physicians are male.

Source: *Statistical Abstract of the United States 1987.*

Going in Style

89 percent of Americans who died are embalmed and put in a casket.

Source: Casket Manufacturers Association.

Fat Females Outweigh Fat Males

30 percent of adult women are obese compared to 25.5 percent of adult men. 1 million Wisconsin adults, or about 28 percent of the population, are obese.

Source: 1988 Behavior Risk Factor Telephone Survey.

Uninsured Mamas

Uninsured women spend an average of 1.9 days in the hospital for routine childbirth, while those with traditional insurance spend 2.3 days.

Source: Healthcare Knowledge Systems, Ann Arbor, Michigan.

Bad News First

63 percent of us say we would rather hear the bad news before the good news. The breakdown by age and sex:

		Men	Women
AGE	21 to 34	70%	83%
"	35 to 44	75%	50%
"	45 to 54	76%	53%
"	55+	10%	70%

Source: Mel Poretz and Barry Sinrod. *The First Really Important Survey of American Habits.* Price Stern Sloan, 1989.

Cover-ups

When company arrives unexpectedly, we are most likely to tidy up by tossing things:

In the closet	68%
Under the bed	23%
In the bathtub	6%
In the freezer	3%

Source: Glass Mates Survey of 1,000 adults.

Moonlighting

A record 7.2 million people—6.2 percent of all workers—have more than one job. Who's working two or more jobs:

Men	4.1 million
Women	3.1 million

Source: Bureau of Labor Statistics.

Beer Drinkers Prefer Alcohol in Their Drinks

Nonalcoholic beer accounted for only 100,000 barrels of the 188 million barrels of beer sold last year. By federal law, nonalcoholic beer must be less than 0.5-percent alcohol.

Source: Miller Brewing Company.

Marriage Odds for Women in Their Thirties

1 in 4 black women and nearly 1 in 10 white women now in their mid- to late thirties will never marry.

Source: Study by Yale sociologist Neil Bennet.

Drugs in the Work Force

1 in 4 U.S. workers has personal knowledge of coworkers using illegal drugs on the job.

1 in 3 workers knows of coworkers using illegal drugs before or after work.

Source: Gallup Poll commissioned by the Institute for a Drug-Free Workplace.

Entering a Sweepstakes

88 percent of people over 21 in the United States say they have entered a sweepstakes. The breakdown:

		Men	Women
AGE	21 to 34	100%	100%
"	35 to 44	100%	95%
"	45 to 54	89%	89%
"	55 +	77%	60%
ALL AGES:		89%	88%

Source: Mel Poretz and Barry Sinrod. *The First Really Important Survey of American Habits*. Price Stern Sloan, 1989.

Stressful Party Givers

Party preparation makes 43 percent of women fretful, but only 29 percent of men. Those with high incomes ($75,000-plus) report most party-giving stress.

Source: Dixie Products.

Fooling Around

60 percent of U.S. husbands, and 40 percent of U.S. wives have had an affair.

Source: Arland Thornton, University of Michigan Institute for Social Research, Ann Arbor.

Size of U.S. Households

The nation's average household size is 2.62 persons. 56 percent of all households are married couples—down from 61 percent in the 1980s.

Source: U.S. Census Bureau.

Heads of Households

17 percent of families are headed by women.
13 percent of white families are headed by a female, 44 percent of black families, and 23 percent of Hispanic families.

Source: Ibid.

Early Retirement

60 percent of those offered early retirement at major companies in the 1980s accepted.

Source: Charles D. Spencer and Associates.

Senior Management

Nearly a third of executives surveyed in 1989 said they wish to retire before age 60.

Retire before age 60	31%
" age 60 to 65	50%
" age 65 to 70	9%
Work as long as possible	10%

Source: Korn/Ferry International.

Saving for Old Age

Overall, Americans save 4 percent of their annual income. Americans in their preretirement years (ages 45 to 64) save about 14 percent on average.

Source: *Money,* April 1989.

Back to Work

One third of surveyed retired senior executives returned to a full-time job within 18 months of retirement.

Reasons for returning	
Job satisfaction, enjoyment or sense of accomplishment	53%
Remain active or avoid boredom	29%
Contribution to society	18%
How they returned	
Employed by another organization	38%
Self-employed	32%
Free-lance or consultant	23%
Founded own company	7%

Source: Russel Reynolds Association: 1985.

Weekly Allowances

American kids receive the following weekly allowances based on these ages:

Ages	Amount
6 to 7	$ 1.98
8 to 9	4.15
10 to 11	7.82
12 to 13	17.44
14 to 15	32.44

Source: Knight Ridder News Service Dispatch Graphic, 1989.

Homeless Youth

There are about 500,000 runaways and "throwaways" yearly under the age of 18. During the summer of 1989 the number of homeless youths on any night was 68,000.

Source: The General Accounting Office.

Adoption Rates

Married or unmarried women in the 20-to-44 age range adopting an unrelated child was 1.3 percent in 1987.

Source: National Center for Health Statistics and the National Institute on Child Health.

Paid Holidays

18 percent of employers grant paid leave on Martin Luther King Jr. Day. 16 percent for Columbus Day. 45 percent for Presidents' Day.

Source: The Bureau of National Affairs.

Trash Piles Up

Each month American households generate 28 lbs. of newspaper, 17.2 lbs. of glass containers, 6.8 lbs. of tin cans, 4 lbs. of cardboard, 1.9 lbs. of aluminum cans, and 0.6 lbs. of plastic.

Source: Browning-Ferris Industry.

New Homes

The average new home is 1,810 square feet. 82 percent of new homes have three or more bedrooms. 65 percent of new homes have a built-in fireplace. 79 percent of new homes come with a garage. $145,500 was the average price of a new home nationwide in 1988.

Source: Congressional Research Service.

Wall Street Illiterates

One half of all consumers—including a quarter of those with incomes above $50,000—say they don't even know how to buy shares of stock.

Source: The Roper Organization and Peter D. Hart Research Associates.

Senior Volunteers Work

In 1987, 40 percent of all Americans 65 to 74 were either volunteering in organizations or informally helping others on a regular basis.

Source: Susan M. Chambre, a Baruch College sociologist.

Older Moms

There were 36,156 live births of babies to women 40 to 49 years old. Women aged 40 to 44 gave birth to 34,781 of those newborns, while women in the 45-to-49 age group accounted for only 1,375 of the births.

Source: The National Center for Health Statistics, 1987.

Father's Day Gifts

59 percent of consumers buy Father's Day gifts, spending an average of $25.14.

Source: Gift and Stationery Business Survey. The Gallup Organization.

Preacher Women

20,730 women are ordained to the whole ministry, which is 7.9 percent of North American clergy.

Source: Yearbook of American and Canadian Churches, 1989.

Municipal Solid Waste

Each year Americans generate over 150 million tons of garbage. 80 percent goes to landfills, 10 percent is incinerated, and only about 10 percent is recycled.

Source: The Garbage Project. University of Arizona, Tucson.

Our Favorite Videos

Comedies	95%
Action/Adventure	92%
Drama	91%
Science Fiction	63%
Film Classics	61%
Children's	49%
Horror	46%
Western	46%
Music Videos	35%
X-Rated	17%
How-to	14%
Others	5%

Source: American Video Association.

Which Charities Do We Support?

The following are percentages of households that give to various types of charities with the average annual household contribution shown in parentheses. (Note that percentages do not add up to 100 percent because many households give to more than one charity.)

Churches, synagogues ($715)	53%
Human services ($210)	24%

Health ($130)	24%
Youth development groups ($88)	19%
Education ($293)	15%
Environmental groups ($87)	11%
Arts, culture, humanities ($260)	8%
Public and social benefit ($153)	7%
Private and community foundations ($145)	5%
International ($281)	4%

Source: National Center on Charitable Statistics.

Heroic Figures to Kids

Actors	38%
Musicians	19%
Athletes	11%
Comedians	11%
Politicians	6%

Source: Louis Harris/Girl Scouts of the United States, poll of 2,000 kids, grades 7–12.

Shy Boys Use More Drugs

Shy boys are more likely to use cocaine and hallucinogens than any other group: 8.8 percent of them used cocaine compared with 2.3 percent of nonshy boys.

Source: Randy M. Page, assistant professor of health education at the University of Idaho, Moscow.

Age 65 and Growing

The elderly population rapidly increases:

| 1986 | 12% of our population |
| 2030 | 21% of population (forecast) |

Source: Family Service America.

Dining in a Dash

One fifth of us eat at a fast-food restaurant each day.

Source: Wendy's Gallup Survey of 1,029 fast-food consumers.

Hispanic Growth

In 1974, 4.5 percent of the U.S. population was Hispanic.

In 1988, 8.1 percent of the U.S. population was Hispanic.

Source: U.S. Population Estimates, 1980–1988. U.S. Census Bureau.

The Big Move to Nevada

Nevada's population was the United States' fastest growing between 1980 and 1989, jumping over 39 percent. Biggest loser: West Virginia with a 5-percent drop.

Source: U.S. Census Bureau.

Unpaid Women

The United States has almost 42 million women aged 16 and older who aren't in the paid labor force. The paid labor force has 56 million women.

Breakdown of activities:
26 million women said they had to tend house.
4 million are going to school.
1.5 million are unable to work.
10 million aren't in the labor force for other reasons, including retirement.

Source: U.S. Bureau of Labor Statistics, 1989.

Drinking Habits

32 percent of all Americans have no more than three drinks per week.

22 percent of all Americans have no more than two drinks a day.

11 percent of all Americans have more than two drinks per day.

Source: Institute of Medicine, an agency of the National Academy of Sciences.

Most Populous Nations

China	1.1 billion
India	833 million
Soviet Union	289 million
United States	250 million

Source: U.S. Census Bureau.

U.S. Population Growth Patterns

With about 250 million people, the United States ranks 4th among all nations, which is 10.4 percent higher than the 1980 census of 226,545,805.

The population reached 200 million in 1967, 150 million in 1949, 100 million in 1915, and 50 million in 1880.

The first census in 1790 counted fewer than 4 million Americans.

Projections show the U.S. population could reach 300 million in about 30 years. Every day the population grows by about 6,300 people with a 4,400 surplus of births over deaths. The rest comes from immigration. The net gain of 1 person every 14 seconds is based on 1 birth every 8 seconds, 1 death every 14 seconds, 1 immigrant every 35 seconds, and 1 person leaving the country every three minutes.

Source: Ibid.

Homeowners vs. Renters

Homeowners are more than twice as likely to have washers, 93 percent vs. 42 percent. 84 percent have dryers compared with only 33 percent of renters.

Over half, 55 percent, have dishwashers compared with 31 percent of renters.

Nearly half of homeowners have a separate dining room compared with 24 percent of renters.

83 percent of homeowners have a porch, deck, patio, or balcony vs. only 61 percent of renters. 72 percent homeowners have a garage or carport compared with 29 percent of renters.

When it comes to producing waste, however, renters are nearly tied with owners: 36 percent of renters vs. 38 percent homeowners.

Source: Ibid.

Age According to Race

	Median Age	Percent Under 35	Percent 35–64	Percent 65 and Older
Hispanics	26.1	68	27	5
Blacks	27.7	63	28	8
Native Americans and Asians	29	61	32	7
Whites	33.6	53	34	13
Total Population	–	54	12	–

Source: U.S. Census Bureau Estimate, 1989.

Childless Households

In 1988 the United States had nearly 92 million households (1 or more persons living in the same dwelling). Nearly 3 in 5 U.S. households contain no children.

More than half (53 percent) of all married couples have no children under 18 living at home.

Source: U.S. Census Bureau.

Single-Parent Households

Households headed by female single parents continue to outnumber those headed by male single parents nearly 4 to 1.

Source: Ibid.

American Families

Families in America now account for 7 in 10 households (72 percent).

Married-couple families constituted 57 percent of all households during 1988.

Source: Ibid.

Nonfamily Households

Nearly 30 percent of all U.S. households are nonfamily households. Among nonfamily households, more than 8 in 10 are 1-person households. There are 2.3 million unmarried couples in the United States, almost one third of whom are raising children.

Source: Ibid.

Marital Status

In 1988 there were 9.7 marriages per 1,000 persons—the lowest rate since 1967. Historically 90 percent of Americans marry. The divorce rate in 1988 was 4.8 divorces per 1,000 persons. Although couples aged 20 to 29 represent only 14 percent of all married couples, they are involved in almost 33 percent of all divorces.

Source: Ibid.

Terms of Endearment

Honey	24%
Sweetheart	4.7%
His/Her Name	6%

Dear	4.6%
Baby	5.4%

Other terms high on the list: Lover, Darling, Sugar, Pumpkin, Angel.

Source: Korbel Champagne Cellars' Department of Romance, Weddings and Entertaining (survey of 1,000 people).

Gay Households

Of the nation's 92 million households, 2.6 million are inhabited by unmarried couples of the opposite sex and 1.6 million households consist of unmarried couples of the same sex. While there are many reasons why unmarried people (domestic partners) share the same households, an estimated 40 percent of the domestic partners are gay.

Source: *Time*, November 20, 1989.

The Odds of Remarrying vs. Never-Marrieds

In 1986, the marriage rate for divorced women was 79.5 per 1,000 versus 59.7 per 1,000 for never-married women. The rate for divorced men was 117.8 per 1,000 versus 49.1 per 1,000 for never-married men.

Source: U.S. Census Bureau.

Adoption Odds

More than 1 million couples are seeking to adopt 30,000 white infants in the United States each year.

Source: National Committee for Adoption.

Fast-Food Consumption

$230 per capita was spent on fast food in the United States during 1988.

Source: *Restaurant Business*.

Choice for Fast-Food Dining

If you're going to eat fast food, the odds are 2 out of 5 times you will buy it at McDonald's. The odds are 1 out of 5 times you will choose Burger King, and 1 out of 10 times for Hardee's and for Wendy's.

Source: Standard and Poor's Industry Survey.

Shaving Habits

2 out of 3 Americans shave with disposable razors.

Source: *The Wall Street Journal*, September 19, 1989.

Sleepwalking

About 5 percent of the adult U.S. population walk in their sleep.

Source: *The Wall Street Journal*, October 3, 1989.

Increase Those Pledges

93 out of 100 United Way campaigns have increases in pledges.

Source: The United Way.

The Firstborn Are Winners

52 percent of U.S. presidents and 21 of our first 23 astronauts were firstborn children.

Source: Leman and Kevin. *Growing Up Firstborn: the Pressure and Privilege of Being Number One*. Delacorte Press, 1989.

Serve Yourself

81 percent of the gasoline sold in the United States is self-served.

Source: Amoco Oil Company.

Chocolate Chips

25 percent of all cookies consumed in the United States are chocolate-chip cookies.

Source: *Gourmet Retailer*, November 1989.

Americans Are Pet Lovers

In 1989, Americans owned 54.6 million cats as compared with 52.4 million dogs. They owned 12.9 million birds.

Source: American Veterinary Medical Association.

Metropolitan and Nonmetropolitan Populations

More than 3 out of 4 Americans (77 percent) now live in a metropolitan area that includes a central city of at least 50,000 and the towns and cities economically tied to it.

One fifth of the nation's metropolitan areas lost population in the 1980s. More than 4 in 10 Americans now live in the suburbs.

Source: U.S. Census Bureau.

American Indians

In 1989 there were 949,075 American Indians, Eskimos, and Aluets living on or near Indian reservations in the United States.

Source: U.S. Bureau of Indian Affairs.

Asians

More than half (53.3 percent) of all Asians employed in the United States hold managerial or professional positions. 34 percent of Asian Americans have a college degree, which is well above the 20-percent average for the nation as a whole, and compared to 11 percent of blacks and 8.6 percent of Hispanics with college sheepskins.

Source: U.S. Bureau of Census.

Blacks

Blacks are the largest minority group in the United States. In 1987 there were 29.8 million blacks, constituting 12 percent of the population. In 1987, 82 percent of blacks were high-school graduates, 11 percent college graduates. 6 percent hold managerial positions in the United States work force. Black families average 3.5 members. In 1987, female-headed families constituted 42 percent of all black families. 56 percent of all blacks live in the south, only 9 percent in the west.

Source: Ibid.

Hispanics

In 1988 there were an estimated 19.4 million Hispanics in the United States. Hispanics constitute 8.1 percent (1 out of every 13 persons) of the U.S. population. 55 percent of all Hispanics live in only two states: Texas and California. Median income of Hispanics in 1987 was $20,306 compared with $31,610 for non-Hispanic families. Nearly 40 percent of Hispanic youths drop out of high school compared with 17 percent blacks and 14 percent whites.

Source: Ibid.

No Weigh

About 37 percent of Americans never weigh themselves. 14 percent use the scales four or more times a week.

Source: Mel Poretz and Barry Sinrod. *The First Really Important Survey of American Habits*. Price Stern Sloan, 1989.

Dirty Work

80 percent of men say they take out the garbage.

Source: Ibid.

Nail Biters

An amazing 92 percent of the U.S. population bite their fingernails. And 25 percent admit to biting their toenails.

Source: Ibid.

Immigration

During the 1980s immigration accounted for one third of the nation's population growth.

Source: U.S. Immigration and Naturalization Service.

Refugees

In 1987 nearly 65,000 refugees arrived in the United States. 35 percent of them were from Vietnam.

Source: U.S. Census Bureau.

Population Migration

Between 1980 and 1987 25 states gained residents by people moving in from other states or from abroad for a net gain of 8 million people. The other 25 states had a net loss of 3.1 million people. California took in the most new residents, followed by Florida, Texas, Arizona, Georgia, North Carolina, and Virginia. One third of the U.S. population lives in 1 of those 7 states.

Source: Ibid.

On the Move

Nearly 1 in 5 Americans (18 percent) changed residences between 1986 and 1987; 11 percent of those merely moved somewhere else in the same county, 4 percent remained in the same state, and 3 percent moved to another state.

Source: Ibid.

Poverty Levels

The poverty rate for Americans over 65 fell from 12.5 percent in 1987 to 12 percent in 1988, the lowest poverty rate for the elderly ever recorded.

The poverty levels in 1988 were $6,024 for an individual and $12,092 for a family of four. A family of nine or more was considered poor if it earned less than $24,133. About 32 million Americans, 13.1 percent, were living in poverty in 1988. The poverty rate among children under 18 was 19.7 percent in 1988. Black children: 44.2 percent; Hispanic children: 37.9 percent; white children: 14.2 percent.

Source: Ibid.

Household Incomes

The median household income in 1988 was $27,230.

Source: Ibid.

Quake Coverage

Only 15 to 20 percent of California homeowners have earthquake insurance, which typically requires a 10-percent deductible and costs between $200 to $400 a year for a $100,000 home.

Source: Association of California Insurance Companies.

Shopping Attitudes

More than a third of people do not like window-shopping or browsing. Complaints about service are so widespread that 6 to 10 people in the survey said they boycotted stores because of the way they were treated.

The percentage was even higher, roughly three-quarters, among individuals earning more than $50,000 per year.

Source: "American Way of Buying." *The Wall Street Journal*, October 13, 1989.

Nukers

8 out of 10 households have microwaves.

Source: "American Way of Buying." *The Wall Street Journal*, September 19, 1989.

Reruns

7 in 10 Americans have VCRs.

Source: Ibid.

More Women Than Men

There are 96 white males for every 100 white females, but only 88 black males for every 100 black females.

Source: U.S. Census Bureau.

Types of American Shoppers

Agreeable shoppers	22%
Practical shoppers	21%
Trendy shoppers	16%
Value shoppers	13%
Top-of-the-line shoppers	10%
Safe shoppers	9%
Status shoppers	5%

Source: Peter D. Hart Research Associates.

A Graying America

In just 30 more years, 49 percent of the population will be over 50 years old. Today, those of us who are 50-plus represent an $800-billion consumer market, control 50 percent of the nation's discretionary income, and own three-quarters of all U.S. financial assets.

In 1987 more than 6 million elderly women lived alone, and nearly three-quarters of them were living in poverty. 4 out of 5 elderly women do not receive any pension income, and the average monthly pension for those who do is $365.

Source: House of Representatives Select Committee on Aging.

Women Are Underpaid

Women with year-round, full-time jobs earned 66 percent as much as men in 1988, up from 60 percent in 1980.

Source: U.S. Census Bureau.

Stepchildren

About 4.5 million families, or almost one fifth of all married couples with children, had at least 1 stepchild under the age of 18 living with them in June 1985. Slightly fewer than half the stepfamilies had a "yours-ours" mix of at least 1 stepchild and at least 1 child born to or adopted by both parents. Slightly more than half had stepchildren only. 9 out of 10 stepchildren lived with their biological mothers and stepfathers; only 740,000 lived with biological fathers and stepmothers. Parents of these stepfamilies are also on the average less well educated. For example, only 35 percent of stepfathers had some college education, compared with 44 percent for all fathers.

Source: Ibid.

Brand Loyalty

Percentage of users of these products who are loyal at one brand:

Cigarettes	71%
Mayonnaise	65%

Toothpaste	61%
Headache remedy	56%
Ketchup	51%
Beer	48%
Automobile	47%
Perfume/after-shave	46%
Pet food	45%
Shampoo	44%
Tuna fish	44%
Gasoline	39%
Underwear	36%
Blue jeans	33%
Canned vegetables	25%
Garbage bags	23%

Source: *The Wall Street Journal* Centennial Survey.

Indian Land

2 percent of America is owned by Indians.

Source: *Statistical Abstract of the United States 1987.* Bureau of the Census. U.S. Department of Commerce.

Man's Best Friend

39 percent of Americans own a dog.

Source: Pet Food Institute Fact Sheet, 1986.

Extramarital Sex

87 percent of American women do not think it is acceptable to have extramarital sex.

Source: "Annual Study of Women's Attitudes." Mark Clements Research Inc., 1987.

Smoking?

97 percent of Americans think smoking is risky.

Source: *Public Opinion,* February/March 1986.

Culture vs. Sports

More people went to the opera (17.7 million) and the symphony (23.3 million) than went to a National Basketball Association game (14 million), a National Football League game (17 million), or a National Hockey League game (12.4 million).

Source: *Statistical Abstract of the United States 1987.*

Total Alcohol Consumption

Total alcohol consumption was about 40.6 gallons per person in 1987.

Source: Ibid.

VCR Owners

The percentage of VCR owners in 1988 was about 66.5 percent.

Source: Ibid.

X-Rated Movie Viewers

In 1988, 31 percent of men had seen an X-rated movie and 17 percent of women had seen one.

Source: University of Chicago Poll, 1989.

America's Poor

More people were living in poverty in 1987 than in 1979 (13.5 percent compared with 11.7 percent).

Source: *Statistical Abstract of the United States 1987.*

Nuking Your Dinner

In 1987, 12,741,000 microwave-oven dinners were bought in the United States.

Source: Ibid.

Foreign-Born Population

The foreign-born population is only about 6 percent in the United States.

Source: U.S. Census Bureau.

Where the Millionaires Live

California leads the nation in the number of millionaires with an estimated 95,000 of them. Florida is second with 41,800.

Source: *Statistical Abstract of the United States 1987*.

Cats and Dogs in the United States

In 1987, cats surpassed dogs as the most popular pet in the United States. There were 54.6 million cats and 52.4 million dogs.

Source: Ibid.

Where Are the Lawyers?

The place with the most lawyers is Washington, D.C., with 1 lawyer for every 22 people. West Virginia has the fewest lawyers per capita with only 1 lawyer for every 689 residents.

Source: Ibid.

Lots of Teachers

The United States has 94,000 barbers, 73,000 professional athletes, 130,000 bill collectors, and 4.4 million teachers.

Source: Ibid.

Going Bananas

The average person eats almost 25 lbs. of bananas a year.

Source: Ibid.

The Youngest and the Oldest Live Far Away

Utah has the youngest population—37 percent are under 18. Florida has the oldest population: 18 percent are 65 and older.

Source: Ibid.

Accidental Children

In 1982, women said that 27 percent of their children were "mistimed" and 8 percent had been unwanted at the time of conception.

Source: Linda Atkinson, Richard Lincoln, and Jacqueline Jarroch Forrest. "Worldwide Trends in Funding for Contraceptive Research and Evaluation." *Family Planning Perspectives*, vol. 17, no. 5 (September/October 1985): 196.

Less Skin

More than 70 percent of the bathing suits sold in the summer of 1990 were one-piece.

Source: NDP Group.

The Way They Go

In 1986 and 1987, 85 percent of the dying chose to be buried and 14 percent chose cremation.

Source: *Family Economics Review*, vol. 2, no. 4.

Watching What You Eat

About 78 percent of Americans worry about the quality and healthfulness of their diet.

Source: The Gallup Organization.

Environmentalism

Over 75 percent of Americans say they are environmentalists.

Source: Ibid.

No Money—No Food

13 percent of Americans said there have been times in the past year when they did not have enough money to buy needed food for the family. This includes 3 percent of college graduates, 11 percent of people that did not complete college, 13 percent of high-school graduates, and 25 percent of people that did not finish high school.

Source: Ibid.

Breakdown of the Workday

Amount of time we work to pay expenses in an 8-hour workday:

Federal and state taxes	2 hours, 45 minutes
Housing	1 hour, 25 minutes
Food, tobacco	57 minutes
Medical care	46 minutes
Transportation	39 minutes
Recreation	25 minutes
Other	1 hour, 3 minutes

Source: The Tax Foundation: 1990.

Teen Spending

America's 13.4 million teenagers ages 12 to 15 spend about $13 a week of their allowances (an annual market of $10.5 billion).

Some of the popular items purchased and the percentage of teens buying them:

Boom boxes	73%
Stereos	63%
Boys' Levi's jeans	62%
Bubble gum	60%
Video games	58%
Nikes	55%

Candy	52%
Girls' Guess? jeans	47%
Reeboks	44%

Source: Teenage Research Unlimited, Northbrook, Illinois.

Americans of Italian Descent Are Not Mafiosi

There are 12,184,000 people living in the United States who are of Italian descent. It is estimated that there are 4,000 Americans who belong to the Mafia.

Source: *Statistical Abstract of the United States 1987; The Today Show*, November 8, 1990.

Pizza Specials

In 1990, 41 percent of Americans who eat pizza when they dine out do so with a special (i.e., a 2-for-1 offer).

Source: NPD Crest.

Mom-and-Pop Shops and Their Kids

33 percent of the nation's small-to-medium-sized business owners report that their children work for them. 75 percent of these entrepreneurs say they want their children to succeed them in the event of death. Less than 50 percent of them have a succession plan.

Source: BDO Seidman.

A Little White Lie

A survey following a presidential election showed that 85 percent of Americans who were asked if they had voted answered yes. Actually, only 50 percent voted.

Source: *TV Guide*.

Where They Vote and Where They Don't

Minnesota had the largest voter turnout in the 1988 presidential election with 66.3 percent of its eligible voters showing up at the voting booths. Montana had 62.41

percent, and in third place was Maine with 62.15 percent. Georgia had the lowest turnout with only 38.79 percent of its eligible citizens voting. 38.91 percent of the South Carolinans voted, and in the District of Columbia, the turnout was 39.44 percent.

Source: U.S. Census Bureau.

Working Moms

In 1987, 55 percent of women with preschoolers worked in the paid labor force.

Source: Ibid.

Working Senior Citizens

An estimated 12.6 percent of U.S. citizens who are age 65 and older will be working during the period 1990 through 1999.

Source: Older Worker Employment Services.

Seasons Greetings

In 1990, Americans purchased 2.3 billion Christmas cards and mailed about 44 cards per family.

Source: Greeting Card Association.

The American Dream

Here's how Americans rank their dreams:

Dream Element	Percent Ranking It Very Important
Having a happy home life	97.8
Giving children a good education	95.7
Competent, affordable health care	91.6
Having a job you like	90.4
Having enough savings	88.7

Dream Element	Percent Ranking It Very Important
Owning a home	82.3
Sending children to a good college	78.7
Living well in retirement	72.4
Being free of debt	71.3
Having enough free time	69.7
Having a job that pays well	69.3
Having children	68.0
Getting ahead on your job	64.4
Being able to work as many years as you want	64.0
Having your home appreciate in value	62.3
Being married	62.3
Living in a nice community	60.3
Being able to travel when you want to	43.3
Having money for occasional luxuries	43.0
Being able to leave inheritance for children	30.0
Retiring early	26.7
Owning your own business	23.0
Owning a late-model car	10.7
Owning a vacation home	7.0

Here's what Americans think are the obstacles that will get in the way of those dreams:

Obstacles	Percent Very Worried
Catastrophic illness in family	41
Environmental problems	39
Cost of education	34

Obstacles	Percent Very Worried
War	34
Rising taxes	33
Stock market collapse	28
Pension cuts	27
Inflation	26
Recession	20
Loss of job	19
Poor investment decisions	19
Cuts in employee benefits	19
Inability to sell house	16
Substance abuse in family	12
Personal debt problems	10
Divorce	10
Natural disasters	8

Source: *Money* and the Gallup Organization.

Teens' Taste for Alcohol

70 percent of high-school students surveyed said they drink alcohol. Of those, how often they drink:

Every day	2%
Few times a week	12%
Almost never	26%
Once a month	29%
Once a week	31%

Source: Chrysler Motors nationwide survey of 2,000 high-school students. *USA Today*, December 10, 1990.

The Second Time Around

14 percent of the U.S. women have been married at least twice. 15 percent of American children who live with two parents live with a stepparent.

Source: 1990 U.S. Census Bureau.

The Divorcée vs. the Widow

About 65 percent of the divorced women remarry in comparison to 23 percent of the widows. Note that the difference is probably due to the average age of the widows, which is higher.

Source: Ibid.

Classified Ad: DW/Child Seeks Mate

69 percent of American divorced woman who have a child under age 18 and are divorced by their first husband remarry. This figure compares to 51 percent of the divorced women without children who remarry.

Source: Ibid.

Classified Ad: WW/Child Seeks Mate

45 percent of widows with a child under age 18 remarry versus only 12 percent of the widows who don't have a child under 18.

Source: Ibid.

A Nation of Gift Buyers

Here's what Americans spend when they buy presents: The average Christmas gift for a close family member or a friend costs $55.50. Wedding gifts come in second place costing $47.90; followed by anniversary gifts at $44.10 and birthday gifts in fourth place at $30.70.

Source: Gallop Poll for *Gift and Stationery Business*.

Mother's Day vs. Father's Day Gifts

The average amount spent per gift for a Mother's Day present is $25.95. For pop on his day, it was $25.14. However, 31 percent of consumers do not purchase a

Father's Day gift as compared to only 19 percent who do not buy gifts on Mother's Day.

Source: Ibid.

Your Card Is in the Mail . . .

Here's the number of cards Americans purchased in 1990:

Occasion	Millions
Christmas	2,200+
Valentine's Day	900
Easter	185
Mother's Day	145
Father's Day	90
Graduation	85
Thanksgiving	40
Halloween	28
St. Patrick's Day	16
Jewish New Year	12

Source: Ibid.

Women in the State House

In 1989, there were 1,273 women in U.S. state legislatures, more than four times the number in 1969. Women accounted for 17 percent of the total state legislators in 1989 as compared to only 4 percent 20 years earlier.

Source: Center for the American Woman and Politics, Rutgers University.

A Nation of House Owners

In 1989, 64 percent of all U.S. households owned their homes. Only 39 percent of the households headed by people age 35 and under owned their own home. Home ownership for those aged 60 to 64 was 80 percent.

Source: U.S. Census Bureau.

Car Owners in the U.S.

1 out of 5 American households own three or more vehicles. More than 50 percent own at least two.

Source: Worldwatch Institute.

BUSINESS

Advertising Dollars vs. Dollars for Education

In 1985, Procter and Gamble, Philip Morris, and R. J. Reynolds/Nabisco companies each spent more on advertising than the U.S. government spent on adult basic education (about $1.1 billion).

Source: *World Almanac 1988*.

Rolls-Royces Roll for a Long Time

It's estimated that more than half of all Rolls-Royce motorcars built since 1904 are still gliding along the road.

Source: Rolls-Royce.

In 30 Minutes or Less . . .

The odds are 9 to 1 that your pizza will be delivered by Domino's in less than 30 minutes.

Source: Domino's Pizza.

Other Odds About Domino's Pizzas

56 percent of all delivered pizzas were by Domino's in 1989.

1 out of 2 customers order pepperoni on their pizza.

8 percent order pizzas with plain cheese only.

3 percent of Domino's customers order its Extravaganza (everything—double cheese, pepperoni, sausage, mushrooms, onions, green peppers, green olives, ham, and ground beef).

Source: Ibid.

Inventions Are Not Always Profitable

Only 3 percent of inventors make a profit on their invention.

Source: National Congress of Inventor Organizations.

Cool Driving

90 percent of the cars made in the United States have factory-installed air-conditioning.

Source: Kidder-Ridder News Service.

It's in the (Overnight) Mail

During the height of its busiest season before Christmas when Federal Express's business picks up from 1.1 million packages and documents to 1.4 million, more than 98 percent arrive on time.

Source: USA Today, December 14, 1989.

What Are the Odds That a Show Will Succeed on Broadway?

The success rate is 1 in 7.

Source: The Wall Street Journal, December 11, 1989.

Women at the Top

A 1990 survey of the 799 public companies on the combined lists of the 1,090 largest U.S. industrial and service companies revealed that of the 4,012 people who were listed as the highest-paid officers and directors of their companies, 19 were women. Less than one half of 1 percent.

Source: *Fortune*, July 30, 1990.

Getting Reimbursed for Business Trips

Percentage of companies that reimburse for these expenses if incurred over a three-day period:

Liquor on airplane	12%
Telephone calls to home	76%
Travel insurance	9%
Excess baggage	30%
Reading material	4%
Laundry	24%
Day-care	3%
In-room hotel movies	13%

Source: 1987 Survey of Business Travelers. Runzheimer International. Sponsored by Murdoch Magazines.

Friendly Takeovers

Hostile takeovers account for only 2 percent of all announced transactions; 25 percent of all deals have occurred in those industries undergoing deregulation: banking, insurance, financial services, communications, and transportation. More than another 25 percent of the total transaction value in 1988 represented foreign purchases.

Sources: The United Way.

New Businesses

During the 1980s, an average of 620,000 new businesses opened up each year.

Source: Small Business Administration, 1989.

For Whom Americans Work

Firms with 250 or more employees employ 56.7 percent of the work force; midsized firms with 100 to 249 employees account for 8.4 percent of the work force; and firms with under 100 employees account for 34.9 percent of the work force.

Source: The United Way.

Family-Owned

14 percent of American families own a business.

Source: *Statistical Abstract of the United States 1987*. Bureau of the Census. U.S. Department of Commerce.

Stockholders

1 out of 5 Americans owns stock.

Source: *Shareownership 1985*. New York Stock Exchange.

Second-Generation Businesses

45 percent of American owners of small businesses are children of owners of small businesses.

Source: *Business Week*, February 8, 1988.

Car Phones

The average monthly cellular phone bill for the 4.4 million car phones in the United States for the first 6 months in 1990 was $89.30. The average length of calls was 2.3 minutes.

Source: Cellular Telecommunications Industry Association.

Fewer Than 500 Employees per Company

62 percent of American employees work in companies with fewer than 500 employees.

Source: *The Economist*, March 5, 1988.

Fewer Than 100 Employees

93 percent of American companies have fewer than 100 employees.

Source: Ibid

We Need More Health-Care Benefits

Nearly 4 out of 5 workers who went on strike in 1989 did so for health-benefits reasons. The average amount spent on health care in 1989 was $2,536.

Source: Service Employees International Union.

No Premium Health Insurance

57 percent of workers who received health-care benefits in 1988 paid nothing in premiums for their own policies; 36 percent paid nothing for family coverage.

Source: U.S. Bureau of Labor Statistics.

Finding a New Job

The average time to find a new job was 5.9 months in 1989.

Source: Drake Beam Morin.

Pretty People Make More

Researchers found that good-looking men and women make more once hired than less attractive people. They also found that tall men earn more than short men and overweight men earn less than men of normal weight.

Source: University of Pittsburgh Graduate School of Business.

Automated Teller Machines

46 percent of American families have an automated-teller-machine card, and of those families 69 percent use the card for withdrawal.

Source: U.S. Department of Commerce.

Successful Entrepreneurs

77 percent of new businesses live to see their third anniversary.

Source: American Express Company and the National Federation of Independent Business.

The Higher the Investment, the More Likely to Succeed

Of new companies whose initial investment exceeded $50,000, 84 percent survived. But of the businesses with an initial investment of less than $20,000, only 74 percent made it.

Source: Ibid.

Industry in Ocean Kills

Ocean industries cause the deaths of thousands of sea animals each day:

Sea Animals	Number Killed	Cause of Death
Sea turtles	30	Shrimp trawl nets
Dolphins	274	Tuna fishermen
Waterfowl	5,479	Lead poisoning

Source: Tom Heymann. *On An Average Day.* Ballantine Books, 1989.

Drugs at Work

Percentage of problems caused by drugs:

Absenteeism	54%
Accidents	30%
Increase in Medical Expenses	30%
Insubordination	30%
Thefts	36%
Product or Service Problems	33%

Source: Hoffman-La Roche Inc., October 1989.

Marital Problems Affect Productivity

57 percent of small-business owners say their workers' productivity has suffered because of personal and emotional problems. 44 percent of those companies surveyed say their companies have no one designated to help troubled workers.

Source: The American Psychological Association and the Ohio Psychological Association.

Coca-Cola Success

Coca-Cola sells roughly 47 percent of all the soda pop consumed worldwide.

Source: *The Wall Street Journal*, December 19, 1989.

Be a Marlboro Man

Philip Morris's Marlboro—the industry's leading smoke—captures 25 percent of the U.S. market share.

Source: Philip Morris.

Drug-Screening Programs

1 in 5 businesses has a drug-screening program.

Source: U.S. Bureau of Labor Statistics.

Employees Agree

86 percent of American employees believe testing at work helps deter drug use.

Source: The Gallup Poll.

This Bud's for You

For every $100 spent on network television advertising, $4.40 goes toward beer commercials.

Source: *Columbus Dispatch*, February 8, 1990.

Direct Deposits

While most Japanese and about 90 percent of European workers get paid by companies that deposit salaries directly into checking accounts, only about 10 percent of the American work force does so. Studies say the productivity boost from cutting costs and saving time would be about $2.5 billion to $3.4 billion a year.

Source: National Automated Clearing House Association.

It's in the Packaging

4 percent of consumer expenditures on goods in the U.S. goes to packaging, or $225 a year per person.

Source: Worldwatch Institute.

Age of CEOs

The typical CEO of America's largest companies takes power at the age of 51. Current chief executives average 56 years old.

Source: *Business Week*, September 25, 1989.

Getting a Patent

The average patent-pending time in 1988 was 19.9 months. Of the 148,183 patent applications filed in 1988, 83,594 were issued. About 45 percent of all applications filed were from foreign residents and of these, about 47 percent of them were issued.

Source: U.S. Department of Commerce Patent and Trademark Office.

Getting a Trademark

In 1988, 76,813 applications were submitted for trademarks, and 52,461 were registered in an average time of about 13 months.

Source: U.S. Department of Commerce Patent and Trademark Office.

Stocks Make the Money

The cumulative returns for stocks from 1926 to 1989 were 21 times greater than long-term corporate-bond returns and more than 55 times greater than treasury-bill total returns.

Source: Ibbotson Associates, Inc.

Job Hunting

Percentage of executives in different salary ranges whose job searches took:

	Up to 6 Months	6 to 12 Months	Over One Year
Up to $50,000	89%	11%	–
$50,000 to $100,000	53%	39%	8%
$100,000 to $150,000	36%	46%	18%
$150,000 to 200,000	38%	50%	12%
Over $200,000	33%	33%	33%

Source: Polson and Company.

Seeking More Money

Of executives switching jobs, two-thirds make more money than their previous jobs and 40 percent had to relocate.

Source: Ibid.

It's Not What You Know; It's Who You Know

79 percent of executives say they use personal contacts to identify job possibilities, while 69 percent use an executive search firm. 54 percent approached a targeted employer directly, and 41 percent read classified ads. (Executives gave multiple responses.)

Source: Ibid.

Looking for a Part-Time Job?

44 percent of the nation's employers use part-time help, and retailers are number 1 on the hiring list. Other heavy employers of part-timers are, in order, service industries, nursing homes, insurance companies, hospitals, banks, employment agencies, and private-sector transportation services.

Source: U.S. Bureau of Labor Statistics.

More Employees, More Success

Of companies starting off with 6 or more employees, 82 percent survive, while only 71 percent of new companies starting with 1 employee saw their third anniversary.

Source: American Express Company and the National Federation of Independent Business.

Survival Plus Growth

Of new businesses who survived their first 3 years of business, only 37 percent of them added employees and about 15 percent reduced their work force.

Source: Ibid.

Minorities and Women Have a Lower Survival Rate

About 22 percent of new entrepreneurs are women. Their survival rate is 71 percent compared with 78 percent for men. Minority-group members account for 6 percent of the sample and 66 percent of them survived 3 years, compared with 78 percent for nonminority owners.

Source: Ibid.

Becoming an Owner

64 percent of business owners started it themselves, 30 percent purchased it, 2 percent inherited it, 1 percent

were promoted to ownership, and 2 percent were brought into ownership.

Source: Ibid.

Manage, Don't Practice

At large firms of over 100, that is. Partners in large law firms earn an average of $217,000 a year and only practice about one third of the day, while about half of the day is spent supervising the firm's day-to-day operations.

Source: Altman and Weil, Inc.

Ivy Leagues Breed CEOs

Of the top 10 undergraduate schools on the corporate elite list, 5 are Ivy. The top 3 are Yale (39 CEOs), Princeton (32), and Harvard (29). Penn (23) and Northwestern (21) round out the top 5. Cornell (20) is sixth.

Source: *Business Week*, October 20, 1989.

Where Are CEOs From

Of the 1,000 most valuable U.S. companies, New York led the list of native states with 136 CEOs born there, followed by Pennsylvania with 71 and Illinois with 67.

Source: Ibid.

Religious CEOs

Of the 1,000 most valuable U.S. companies, 62 percent of the CEOs are Protestants, 23 percent are Roman Catholics, and about 13 percent are Jewish.

Source: Ibid.

Pension Plans

46 percent of full-time workers participated in employer-financed pension plans in 1988. 49 percent of males and 43 percent of females were covered.

Source: *Social Security Bulletin*.

IRS Losing Track of You

In 1988, 72,000 taxpayers did not receive their tax-refund checks averaging about $555 each because they were undeliverable.

Source: U.S. Treasury Department.

Average Amount of Life Insurance

The average amount of life insurance is $108,000 per household with life insurance.

Source: American Council of Life Insurance.

Investment Bankers Hurt the Economy

68 percent of CEOs think that the deals driven by investment bankers hurt America's ability to compete on the global market.

Source: Clark Martire and Bartolomeo.

Reasons for Being Fired

Incompetence	39%
Inability to get along with others	17%
Dishonesty or lying	12%
Negative attitude	10%
Lack of motivation	7%
Failure to follow instructions	7%
Other reasons	8%

Source: Robert Half International, Inc.

High Anxieties

Executives' biggest anxieties:

Loss of job due to merger or acquisition	54%
Burnout	26%
Failure to get promoted	8%
Being fired	6%

Failure to get a raise	5%
Insufficient income to meet living standards or financial obligations	3%
Illness	2%

Source: Robert Half International, Inc.

Where CPAs Work

In 1989, an estimated 46 percent of CPAs worked for public accounting, about 40 percent worked in business and industry, and the rest worked for education and government.

Source: American Institute of Certified Public Accountants.

Junk Mail

About 67 percent of consumers open direct mail; 30 percent open some and throw the rest away; and 3 percent throw all away without opening.

Source: Simmons Market Research Bureau, 1984.

Buying Through Direct Response

More people would buy through direct response if:

They gave a money-back guarantee	82%
Had a toll-free number	78%
Free trial period	75%
They sent a receipt or confirmation of order	72%
Able to order by phone	66%
Billed after order was received	63%
Pay over period of time	47%
Gave charge account	39%
Could use credit card	34%

Source: Joseph Castelli and François Christen. *VALS Look at Direct Marketing.* 1986.

Being Audited

The odds of being audited by the IRS are 1.03 percent for all individuals in 1988 and about 2.32 percent of individuals with an income of $50,000 or more.

Another 3.5 percent of taxpayers receive computer notices, so the odds jump to 4.5-percent chance of being contacted by the IRS.

Source: National Taxpayers Union.

Odds of Getting a Refund

About 75 percent of all tax returns filed result in a refund.

Source: Ibid.

IRS Telephone Advice

When you call the IRS free tax assistance hot lines, the chances of you getting a wrong answer range between 35 to 40 percent of the time.

Source: Ibid.

Using a Professional

47 percent of the tax returns filed by individuals are prepared by tax professionals.

Source: Ibid.

Serving Time for Tax Evasion

1,590 taxpayers received a prison sentence for criminal tax violations. Assuming that roughly 100 million taxpayers file tax returns, the odds of a taxpayer receiving a sentence are .0016 percent.

Source: Ibid.

Jobs in the Future

The number of jobs will increase by only about 15 percent in the 1990s, about half as fast as in the 1980s. Paralegals and health-related technicians will experience the largest growth rate.

Source: U.S. Bureau of Labor Statistics.

Unemployment Rates

5.3 percent of people over age 16 were out of work in 1989. For whites it was 4.5 percent, blacks 11.4 percent, and Hispanics 8 percent.

Source: U.S. Department of Labor, Bureau of Labor Statistics, February 2, 1990.

A Woman's Best Friend

DeBeers, a London based company, sorts and values about 80 percent of the world's rough diamonds before selling them to wholesalers.

Source: DeBeers Consolidated Mines Ltd.

Industrial Costs

Wages and benefits account for about 75 percent of industrial costs.

Source: Gordon Richards, an economist for the National Association of Manufacturers.

Junk-Bond Owners

The size of the junk-bond market is about $200 billion in face value outstanding. Who owns these bonds:

Insurance companies	30%
Mutual funds, money managers	30%
Pension funds	15%
Foreign investors	9%
Savings and loans	7%

Individuals	5%
Corporations	3%
Securities dealers	1%

Source: Drexel Burnham Lambert.

Corporate Takeovers Terminate Companies

Corporate takeovers have resulted in more than 2,000 company plants being terminated in the 1980s.

Source: American Association of Retired Workers.

Discrimination Against Women

46 percent of women feel they have experienced discrimination at work.

Source: Wick and Company.

Quitting Jobs at Fortune 500 Companies

Over half of all men and women who quit their jobs say it had nothing to do with their personal life. About 26 percent say they quit for their spouse. About 10 percent quit for financial reasons. 26 percent of men and only 9 percent of women quit for children.

Source: Ibid.

Women Secretaries

98 percent of America's 5.7 million office workers are women.

Source: 9 to 5 Organization for Women Office Workers.

Busy Business People

Business people can only be reached by a single phone call about 17 percent of the time.

Source: *American Demographics* magazine.

The Big Board

The New York Stock Exchange handled 69.23 percent of the trades of its listed stocks in 1989.

Source: *The Wall Street Journal*, April 17, 1990.

Complaints Against Lawyers

In 1988, about 93,000 complaints were filed against lawyers; about 40 percent of the clients were complaining about alcohol or drug dependencies.

Source: HALT, a Washington D.C. legal reform group.

More Than Our Share

The United States, with 5 percent of the world's population, consumes 25 percent of its energy supply.

Source: American Petroleum Institute.

Gas Guzzlers

Cars and trucks require 45 percent of the oil used in the United States.

Source: Ibid.

A Barrel of Oil in the United States

Out of 1 barrel of oil, here's where it pours:

Gasoline	46%
Diesel, heating oil	21%
Jet fuel	10%
Residual fuel oil	7%
Natural gas	5%
Coke production	4%
Asphalt and road oil	3%
Petrochemical	3%
Feed stocks	3%
Propane	3%

| Lubricants | 1% |
| Kerosene, waxes, other | 1% |

Figures add up to more than 100 percent because of the rounding and a volume gain in the processing crude.

Source: American Petroleum Institute and U.S. Department of Energy.

We Wish You a Merry Christmas

An estimated 79 percent of businesses with sales between $2 million and $100 million provide for a holiday office party and 51 percent give gifts.

Source: BDO Seidman.

TRAVEL

Airline Accidents

4 out of 25 commercial airline accidents in the U.S. are caused by traffic control.

Source: *The New York Times*, March 20, 1988.

Close to Work

18 percent of American workers live within 10 minutes of work.

52 percent live within 20 minutes of work.

Source: *1980 Census of Population: General Social and Economic Characteristics, United States Summary.* Bureau of the Census. U.S. Department of Commerce, 1983.

Japanese Tourists

18 percent of all visitors who came to the U.S. are Japanese.

Source: Daniel Evan Weiss. *100% American.* Poseidon Press, 1988.

Carpooling

1 out of 5 Americans carpools to work.

Source: *1980 Census of Population: General Social and Economic Characteristics, United States Summary.*

Visiting Yellowstone

30 percent of all Americans have visited Yellowstone National Park.

Source: Lewis Lapham, Michael Pollan, and Eric Etheridge. *The Harper's Index*. Henry Holt, 1987.

Going Abroad Alone

31 percent of all Americans who travel overseas travel alone.

Source: Somerset Waters. *The Big Picture—1987*. Travel Industry World Yearbook. Child and Waters, Inc.

Female Travelers

53 percent of all Americans who are issued passports are female.

Source: Ibid.

Flying High

72 percent of Americans have flown in commercial airlines.

Source: "The Bristol Myers Report: Medicine in the Next Century."

Odds of Going to Atlantic City

While the odds of winning at the gaming tables in Atlantic City aren't in your favor, the odds of going there are. More people will visit this New Jersey gambling community than any other place in the United States. 33 million people visit Atlantic City each year, or 1 out of 7 Americans.

Source: *Time*, September 25, 1989.

Arriving on Schedule

79.6 percent of flights by the largest U.S. carriers arrived on time during the month of September 1989—that is, within 15 minutes of schedule.

Source: U.S. Department of Transportation.

What Are the Odds That You'll Get the Same Flight Number As a Plane That Crashed?

Zero. The U.S. airlines eliminate the flight number after a crash.

Source: *The Wall Street Journal,* December 11, 1989.

Americans Go Cruising

In 1989, about 3.5 million Americans took cruises.

Source: Cruise Line Industry Association.

Welcome to America

38.5 million foreigners visited the United States in 1989.

Source: U.S. Travel and Tourism Administration.

Overnight Business Trips

While the number of business trips for Americans was 155.6 million in 1983, only 67 percent involved hotel stays, off from 72 percent in 1982 when 103.6 million traveled for business reasons.

Source: Hotel and Travel Index.

Wide Railroad Tracks

One track of double-track railroad can handle as much freight as a 10-lane highway.

Source: Association of American Railroads.

Cruisin'

An estimated 5 percent of all Americans go on a cruise each year. Of these tourists, 33 percent visit the Caribbean. Evidently Americans love cruises—9 out of 10 go cruisin' again.

Source: Cruiseline Industry Association.

Still Putting Along

About 25 million Model-T Fords were made from 1908 to 1927, and roughly 450,000 are still in running condition. Of the 4 million Model-A's made from 1928 to 1931, at least 160,000 survive.

Source: *The Wall Street Journal,* March 20, 1990.

Car Operating Costs

Depending on the size of car you drive, it will cost you between 33 to 44 cents per mile to own and operate a 1990 car, which comes to about $3,300 to $4,400 a year, or $9.04 to $12.05 per day. (Costs include gas, oil, service, maintenance, and tires.)

Source: American Automobile Association.

Radiation from Flying

10 radiation cancer deaths are predicted for every 1,000 crew members flying New York to Seattle for 960 hours a year for 20 years. Frequent fliers on the same route flying 480 hours annually over 20 years have half the risk—5 per 1,000.

Source: U.S. Department of Transportation.

Americans Drive

The average American drives about 10,000 miles a year.

Source: American Automobile Association.

College Spring-Break Locations

Daytona Beach, Florida, is the "hot spot" for college spring breaks; in 1990, it brought in about 400,000. Second is South Padre Island, Texas, drawing 225,000 college students. Then Cancún, Mexico, with 110,000, and Key West with 40,000. In 1985, Fort Lauderdale drew a crowd of 350,000, but in 1990 only 20,000.

Source: USA Today Research, hotel/tourist bureaus in each city.

Where We Drive

To work	34.3%
Family personal business	30.4%
Social activities, recreation	30.0%
Civic, education, religious	4.1%
Other	1.2%

Source: *1990 Highway Fact Book*. Highway Users Federation Survey.

Auto Premiums

The average annual cost of an auto-insurance premium in the United States was $517.71 in 1988.

The states with the lowest premiums are:

Alabama	$278.33
Iowa	292.51
South Dakota	324.90
Tennessee	338.46
North Dakota	343.85

Source: A. M. Best Company.

Night Driving

Who has some problems with driving in the dark:

Men	34%
Women	58%
Overall	45%

Source: Eveready Battery Company.

Busy Airport

Chicago's O'Hare Airport is America's busiest airport. In 1988, 28.9 million passengers visited O'Hare.

Source: Federal Aviation Administration.

Mother Nature Traffic Delays

Weather accounts for 70 percent of air-traffic-control delays.

Source: Ibid.

Unhealthy Travel for Women

While on the road, women are more likely to skip breakfast than men and half as likely to order an entrée salad for dinner. Women are also five times more likely to order room service than men.

Source: Runzheimer International Study of Business Travel.

Coast to Coast on Greyhound

It takes about three days to cross the country by bus, costing only $68. An airplane only takes about three hours but costs considerably more. ($350, TWA coach, 3 days in advance, L.A.–N.Y., one-way.)

Source: Greyhound Lines Inc., USA Today Research.

Typical Greyhound Riders

The typical Greyhound rider is a young white female traveling alone.

Source: Ibid.

Arriving on Time

From September 1987 to April 1989 the most on-time airlines were American West, Southwest, and American, arriving as scheduled 81 to 85 percent of the time.

Source: U.S. Department of Transportation.

Check for Yourselves

About 37 percent of travel agents say they try to talk their customers into buying a certain airline ticket even if it isn't the cheapest, just so they can get a special prize.

Source: *Travel Life*.

Old Planes

The average age of the nation's 3,671 commercial airliners is 13 years.

Source: Federal Aviation Administration.

Built to Last

A Boeing 747 is designed to fly 60,000 hours and to undergo 20,000 takeoff and landing cycles in 20 years. A short-hop 737 is designed for 45,000 hours in the air and 75,000 takeoffs and landing cycles in 20 years.

Souce: Ibid.

Surviving a Crash

About 56 percent of the people involved in an airplane crash survive.

Source: John Enders, president of Flight Safety Foundation of Arlington, Virginia.

Flying on a Plane

Three-quarters of Americans have flown at least once, and 30 percent of these travelers fly three times a year.

Source: Peter D. Hart Research and Associates.

Visiting the United States

In 1989, almost 39 million tourists visited the United States.

Source: U.S. Travel and Tourism Administration.

Most-Visited Caribbean Islands

Americans visited the Caribbean Islands a lot in 1988.
1,426,000 people visited Puerto Rico
1,275,000 visited the Bahamas
550,000 visited the U.S. Virgin Islands
460,000 visited Jamaica
400,000 visited the Dominican Republic

Source: Caribbean Tourism Organization.

Winter Getaways

In 1988, 3.5 million people went on a cruise, 54 million people visited ski resorts, and 40 million people went to Florida.

Source: Julie Bowers of the U.S. Travel Data Center.

Traveling over Thanksgiving

Over 31 million of us travel over the Thanksgiving holiday.

Source: American Automobile Association.

Airbags on the Rise

In 1990, about one third of all new cars to be sold in the United States have air bags.

Source: *Public Citizen*. The Center for Auto Safety.

Mishandled Baggage

In August 1989, Eastern lost the most baggage, at a rate of 21.86 per 1,000. Southwest had the fewest reports of lost baggage, at a rate of 4.23 per 1,000. The average airline mishandled 8.19 pieces of luggage per 1,000.

Source: Air Travel Consumer Report.

Airline Complaints

U.S. airlines had a total of 942 complaints in September 1989; 435 concerned flight problems, 126 baggage problems, and 118 customer-service problems.

Source: Consumer Affairs Division, U.S. Department of Transportation.

Single Cyclers

In 1985, about 48 percent of motorcycle owners were single.

Source: 1985 Survey of Motorcycle Ownership and Usage.

Male Riders

About 91 percent of all motorcycle riders were male in 1985.

Source: Ibid.

Educated Riders

About half of all motorcycle riders in 1985 only finished high school and about two-fifths had achieved some college or postgraduate education.

Source: Ibid.

Experienced Riders

The average motorcycle owner has been riding for 8 years and has owned 3 cycles.

Source: Ibid.

Frequent Trips

About 50 percent of travelers take frequent trips of only 2 to 4 days.

Source: Marketing Projects Group.

Economy Rather Than Luxury

Travelers stay at economy hotels 53 percent of the time and luxury hotels 25 percent. When flying, they choose economy rather than first class (70 percent vs. 5 percent).

Source: Ibid.

Road Trips Rather Than Flying

People are more likely to drive than fly when taking a trip (16 percent vs. 26 percent).

Source: Ibid.

Independent Travel

Travelers go independently rather than on an organized trip (71 percent vs. 12 percent).

Source: Ibid.

Planning Ahead

50 percent of trip goers plan ahead vs. 30 percent who go on the spur of the moment.

Source: Ibid.

Enjoying the Daytime

Daytime activities while vacationing are more important than the nightlife to 62 percent of vacationers versus 14 percent. The remaining 24 percent have no preference.

Source: Ibid.

Dream Trip Overseas

The 2 most desirable destinations overseas are Australia/New Zealand (14.3 percent) and then Great Britain (13 percent).

Source: Ibid.

Need Some Help?

The odds of an AAA member needing a service call is 7 out of 10. The most common car failure is electrical problems—battery, alternator, generator, etc.

Source: American Automobile Association.

Waiting for Help

The average time lapse between a reported call and the arrival of an AAA service person is 25 minutes.

Source: Ibid.

Fuzzbusters

In 1987, 1,834,000 radar detectors were sold in the United States, up from 710,000 in 1983.

Source: *Statistical Abstract of the United States 1987*. Bureau of the Census. U.S. Department of Commerce.

Trains Save Gas

On the average, a truck can move 1 ton of freight 92 miles for each gallon of fuel consumed; a locomotive can move that ton of freight an average of 288 miles on a gallon of fuel.

Source: Federal Railroad Administration.

More Cars Than People in Wyoming

Motor vehicles outnumber people 493,000 to 475,000.

Source: *Statistical Abstract of the United States 1987*.

Extras in Cars Aren't Extras Anymore

	1988 Car Models	
Extras	Domestic	Import
Tilt wheel	74%	49%
Radio/cassette	48%	38%

Extras	1988 Car Models Domestic	Import
Air-conditioning	90%	50%
Power windows	50%	39%
Power door locks	57%	39%
Adjustable seats	28.5%	12.3%

Source: *The Wall Street Journal*, November 9, 1989.

Railroads Save Land

The rights-of-way for streets and roads use 10 times as much land as do the nation's railroad rights-of-way.

Source: Federal Railroad Administration.

Home Sick

Men are more likely than women to miss home while on a long trip.

Source: Marriot Corp.'s Residence Inn.

Designated Drivers

62 percent of American partygoers choose a designated driver most or all of the time.

Source: The Gallup Organization.

Cruising . . .

Look who goes on cruises for their vacations:
50% or more are on their first cruise.
70% are married.
30% are single.
25% take their children along.
50% are under age 48.
72% have annual incomes less than $60,000.

Source: Cruise Lines International Association.

SPORTS

Four Holes-in-One on the Same PGA Hole

On June 16, 1989, 4 holes-in-one were recorded in the U.S. Open in Rochester, New York's Oak Hill Country Club. The odds of 4 PGA Tour players doing it on the same hole on the same day are 1 in 332,000.

Source: *Golf Digest*.

Three-Pointers

38.4 percent of the attempted shots were made from the three-point range during the 1986–1987 season in college basketball.

Source: NCAA.

The On-Side Kick

The odds of recovering an on-side kick by the kickoff team are 15 percent.

Source: An NBC announcer during the December 1, 1989 football game between Alabama and Auburn.

Ties Are Like Kissing Your Sister

Of the 10,080 games played during the regular National Hockey League seasons for the 6-year period ending in 1989, 870 games went into overtime, which is 1 in 12. Of the games that went into overtime, 556 ended in a tie (63 percent). Based on these numbers, the odds of a hockey game ending in a tie is 1 in 20.

Source: National Hockey League.

On-Track vs. Off-Track Betting

88 percent of the bets on horses made in the United States were on-track wagers; the remaining 12 percent were off-track wagers.

Source: Thoroughbred Racing Association.

Average Winning Ticket on a Horse Race

In 1989, the average amount of money won on a winning ticket was $8.80.

Source: Ibid.

Americans on the Run

In 1988, there were between 31.7 and 34.2 million runners in the United States.

Source: American Sports Data, Inc.

America's Skiers

In 1989, there were 21 million skiers in the United States.

Source: National Sporting Goods Association Sports Participation Study.

America's Boats

In 1988, there were 10,362,613 boats in the United States, or about 1 for every 24 persons.

Souce: U.S. Department of Transportation, U.S. Coast Guard.

The Average High-School Coach

The average high-school coach is a 36-year-old white male with 12 years coaching experience. Almost 70 percent of them participated in varsity athletics in college and about 97 percent in high school.

Source: *USA Today*.

Coaches' Biggest Threats

88 percent of high-school coaches say that their biggest threat is booze.

Source: A *USA Today* survey.

Failing the Prop-48 Test

Almost 1 in 15 college freshman athletes fails the Prop-48 Test mostly because they fail to meet the minimum score of 700 on the SAT or 15 on the ACT.

Source: NCAA.

Instant Replay Not So Instant

The average delay per game in the NFL due to the instant replay was 1 minute and 55 seconds.

Source: NFL officials.

Big-Leaguers Make Big Bucks

The average major-league salary is $491,199. The New York Mets average in 1989 was $858,575!

Source: A study by *The New York Times*.

Working Out

1 out of 10 Americans is a member of a health club or fitness center.

13 percent of Americans jog.

Source: George Gallup Jr. *The Gallup Poll—Public Opinion 1986.* Scholarly Resources, Inc.

Bowling

1 out of 4 Americans bowls.

Source: *Statistical Abstract of the U.S. 1987*. Bureau of the Census. U.S. Department of Commerce.

Couch Potatoes

1 out of 4 Americans never exercises at all.

Source: Louis Harris and Associates, Inc. "The Nuprin Pain Report." Bristol-Myers Products: 1985.

Winning Opener Is Advantageous

According to previous finals, NBA teams winning the opening games of the finals have a 70 percent chance of winning the finals.

Source: National Basketball Association.

Pool Sharks

An estimated 10 million different people play pool each year.

Source: *Billiards Digest*.

Cross-Country Skiers

Less than 4 percent of the population cross-country-skis, and they only ski an average of 7.2 days each year.

Source: American Sports Data.

Alpine Skiers

About 7 percent of the population skis downhill. They ski an average of about 7 days a season.

Source: Ibid.

Someone *Should* Bet on the Bay

Derby winners by color of horse since 1875:

Roan	1
Gray	4
Chestnut	36
Bay	53
Dark bay	17
Black	4

Source: Churchill Downs.

Gate 1 Wins Most

How many winners emerged from each gate in Derby since 1900:

Gate	Winners	Gate	Winners
1	12	11	3
2	9	12	3
3	7	13	3
4	10	14	2
5	8	15	1
6	5	16	0
7	7	17	0
8	6	18	0
9	4	19	1
10	8	20	1

Source: Ibid.

Softball Players

There are approximately 41 million amateur softball players in America.

Source: *Statistical Abstract of the United States 1987*. Bureau of the Census. U.S. Department of Commerce.

Snowboarding

Only about .1 percent of the population snowboards, and about 42 percent of these are skateboarders.

Source: American Sports Data.

Average NBA Career

On opening day of the 1989–1990 season, the 351 players on the roster averaged 3.95 years of experience.

Source: Bill Kreifeldt, Clippers PR Director.

Less Than 6 Feet, 7 Feet, or More

11 of the 351 players on the NBA roster in the 1989–1990 season were under 6 feet, and 40 of them were 7 feet or more.

Source: National Basketball Association.

Going Overtime

In the 1989–1990 NBA season, there were 56 single-overtime games, 2 double-overtime games, 1 triple-overtime game, and 1 quintuple overtime game. 1,107 games are played in the season.

Source: Ibid.

Scoring 30 Points

Scoring 30 points or more by 1 player happened 671 times in the 1989–1990 NBA season.

Source: Ibid.

Disqualification

In the 1989–1990 NBA season, 19 players were disqualified.

Source: Ibid.

Hole-in-One on Par-3

The odds for a basic amateur making a hole-in-one on a par-3 hole are about 11,000 to 1. The odds for a PGA pro or an LPGA pro making a hole-in-one on any given round are about 750 to 1 and 3,000 to 1 for an ace on a given hole.

Source: Doctors Francis Shield and Clyne Soley.

Slim Chance for NBA Finals

The odds against a team coming back from a 3-0 or 3-1 deficit in the best-of-7 series to win the finals is very small, since no NBA team has ever accomplished this feat.

Source: National Basketball Association.

Never Learning to Swim

About one third of the population over the age of 14 can't swim.

Source: American Red Cross.

Winning at the Last Minute

The odds against a hockey team winning in the last 5 minutes are 20 to 1 and in the last 2 minutes are 40 to 1.

Source: Sports Products, Inc., Norwalk, Connecticut.

Penalty Shots

Of the 10,080 games played during the regular season in the National Hockey League for the 6-year period ending 1989, 106 penalty shots were taken and only 45 were made. Based on these figures, the chances of making a penalty shot in the NHL is about 42 percent.

Source: National Hockey League.

Home-Court Advantage

Home (field, court, or ice) advantage by sport:

Basketball	.670
Hockey	.581
Football	.574
Baseball	.550

(Figured from the 1989–1990 season)

Men Getting Hurt at the College Level

Odds of getting hurt in various sports at the college level:

Sport	Odds for a Man
Wrestling	2.4 to 1
Football (fall)	3.2 to 1
Ice hockey	3.7 to 1
Basketball	5 to 1
Volleyball	5 to 1
Tennis	7 to 1
Soccer	7.2 to 1
Football (spring)	8.5 to 1
Cross-country	9.3 to 1
Gymnastics	9.6 to 1
Lacrosse	10 to 1
Indoor track and field	10.2 to 1
Outdoor track and field	11 to 1
Baseball	13.7 to 1
Swimming and diving	44.4 to 1

Source: National Athletic Injury/Illness Reporting System, Pennsylvania State University.

Short Football Careers

The average career in football is just over 3 years.
Source: NFL Players Association.

Women Getting Hurt at the College Level

Sport	Odds for a Woman
Gymnastics	2.4 to 1
Badminton	5 to 1
Basketball	6.4 to 1
Cross-country	6.4 to 1
Squash	7 to 1
Indoor track and field	9.6 to 1
Volleyball	9.7 to 1
Tennis	17.2 to 1
Softball	18.2 to 1
Outdoor track and field	19.8 to 1
Field hockey	23.4 to 1
Lacrosse	23.4 to 1
Swimming and diving	39 to 1

Source: National Safety Council.

Decisions for the Heavyweight Title Since 1892

		Percentage of Total
KO	74	37%
TKO	65	32%
Unanimous Decision	36	18%
Others	27	13%

Source: *1990 Sports Almanac*.

The NBA Norm

The average NBA player is 1989 is 6'9", 215 lbs. The 27-year-old has been playing ball in the NBA an average of 4 years.

Source: National Basketball Association.

Daytona-500 Winnings

Since 1959 Ford has won the race the most times (7), followed by Chevrolet with 6 wins and Plymouth with 4 wins.

Source: *1990 Sports Almanac.*

By a Point or 2

From 1982 to 1989, 5 NBA championship games were decided by 1 or 2 points.

Source: Ibid.

USA Breeds Golfers

Major golf tournaments won by golfers from the United States in the eighties:

PGA	10
U.S. Open	9
Masters	5
British Open	5

Source: Ibid.

Making It to the Pros

More than 5 million kids participate in high-school sports at the varsity and junior-varsity levels. However, only 1 in 50 goes on to make a college team, and 1 in 1,000 makes it to the pros. Fewer than 3,500 people in the United States earn a living playing pro sports.

Source: National Federation of State High School Associations.

NCAA Tournament Attendance

The average NCAA basketball tournament attendance in 1989 was 18,586 people.

Source: NCAA.

Average Turnovers

From 1985 to 1989, NFL teams averaged about 36 turnovers each season.

Source: Elias Sports Bureau.

Better to Score First

NFL teams' 1989 record when they score first in a game: 116 wins, 65 losses, 1 tie.

Source: NFL; USA Today Research.

Odds of Home Runs

The odds of a game having 1 home run is 7 to 4; 3 or more 7 to 1 against; 5 or more 100 to 1 against.

Source: Sports Products, Inc., Norwalk, Connecticut.

Reasons to Attend Sporting Events

Get together with friends	59%
Excitement of sports	58%
Loyalty to team	46%
Own participation in sport	30%
Business entertaining	11%

Source: Team Marketing Report.

Packed Games

In the 1980s, the average NFL game was attended by about 60,000 people.

Source: National Football League.

Amazing Games

The odds of seeing a triple play in a game are 1,400 to 1; a shut out 11 to 2; and a starter pitching a no-hitter 1,300 to 1.

Source: Sports Products, Inc., Norwalk, Connecticut.

Right-Handed Tennis Pros

Of the 312 tennis pros listed by the American Tennis Professionals, 267 play right-handed, 43 are lefties, and 2 are ambidextrous.

Source: 1990 ATP Players Guide.

Most Bowl Appearances

Conferences with the most bowl appearances in the eighties:

Southwestern	56
Big Ten	45
Southwest	35
Pacific	35
Big Eight	31

Source: Southeastern Conference.

Being Inducted into the Baseball Hall of Fame

The odds of being inducted into the baseball hall of fame are 1,500 to 1.

Source: Sports Products, Inc., Norwalk, Connecticut.

Draft Picks Making the NFL

The odds for a draft pick making an NFL team are:

Draft Round	Odds
1st	100 to 1
2nd	10 to 1
3rd to 4th	5 to 1
5th to 7th	5 to 4
8th to 10th	4 to 3 against
11th to 14th	7 to 3 against
15th or lower	3 to 1 against

Source: Ibid.

Odds Against a Poor Performance

The odds of a pro football team winning:

While losing in the third quarter	5.5 to 1
While losing with two minutes to play	8.5 to 1
After scoring fewer touchdowns	32 to 1
After gaining fewer yards	3 to 1
After losing more turnovers	4.5 to 1
After making fewer first downs	12 to 1
After running fewer plays from scrimmage	9 to 5

Source: Ibid.

More Than 1 Championship

Only 5 NBA franchises have won more than 1 championship:

Boston Celtics	16
L.A. Lakers	11
Golden State Warriors	3
N.Y. Knicks	2
Philadelphia 76ers	2

Source: USA Today Research.

Payoffs on Recovered Fumbles

The odds for a fumble recovered and returned for a touchdown are 7 to 1 in a single game.

Source: Sports Products, Inc., Norwalk, Connecticut.

Costly Interception

The odds that an interception is returned for a touchdown are 7 to 1 in a single game.

Source: Ibid.

Missing the Extra Point

The odds for at least 1 extra point being missed in a game are 11 to 4.

Source: Ibid.

Punt Returns

The odds for a punt returned for a touchdown are 20 to 1.

Source: Ibid.

Kickoff Returns

The odds that a kickoff will be returned for a touchdown are 45 to 1.

Source: Ibid.

Draftees Making a Team

The NFL drafts 336 players in 12 rounds but only about half ever make a team.

Source: Cordner Nelson. *Careers in Pro Sports*. Rosen Publishing Group, 1990.

U.S. Ski Instructors

There are about 16,000 ski instructors in the United States with about 4,000 of them working full-time during the season. They earn from $1,000 to $4,000 a month for 4 months.

Source: Ibid.

High-School Athletic Injuries

Of the 5.8 million high-school athletes, an estimated 1 million are injured each year, nearly 1 out of 6!

Source: Ibid.

Have Clubs, Will Travel

It costs more than the price of golf clubs to travel the pro golf circuit. A pro golfer must win at least $25,000 to cover the minimum of expenses.

Source: Ibid.

Pros Without Degrees

Only a little over one third of NFL and NBA players have graduated from college.

Source: Associated Press.

Making the NBA Roster

The basketball teams draft about 230 players in June and July, yet only 50 or fewer players make the rosters each year.

Source: Ibid.

Average NFL Play

The National Football League has 1,372 players who average about $250,000 per year.

Source: National Football League.

Free Throws

In the 1989–1990 NBA season 77 percent of the free throws were made.

Source: National Basketball Association.

Indoor Soccer

The average major-indoor-league soccer team played about 52 games during the 1989–1990 season.

Source: Major Indoor League Soccer League.

U.S. Breeds Women Tennis Players

Countries with most top-100 women's tennis players:

United States	31
Italy	8
Australia	7
W. Germany	7
Czechoslovakia	6

Source: Women's International Tennis Association.

Winning Both Games of a Doubleheader

The team that wins the first game has an edge of 9 to 8 for winning the next game.

Source: Sports Products, Inc., Norwalk, Connecticut.

Shooting for Three

In the 1989–1990 NBA season about one third of the attempted three-pointers were made.

Source: National Basketball Association.

Black Athletes in Big-time College Sports

The graduation rate for black athletes in big-time college sports is 26.1 percent in football and 17.1 percent in basketball.

Source: "Win or Lose on Field, Athletes Lose Out Big-Time in Education," Randy Harvey, *Los Angeles Times* (syndicated article in *The Columbus Dispath*, March 17, 1991).

Making the Pros

Of major-college football and basketball players, only about 1 percent play professional sports.

Source: Ibid.

In Baseball, It *Pays* to Arbitrate

In 1991, of the 157 players who filed for arbitration, only a single player took a pay cut.

Source: *The New York Times*.

Baseball Players Make Big Bucks

Of the 650 players being on opening-day rosters, 35 percent receive salaries in excess of $1 million. Of these, 5 percent receive $3 million or more; 14 percent receive between $2 million and $3 million; 16 percent receive between $1 million and $2 million.

Source: Ibid.

NBA Giants

The average height of the 354 players of the National Basketball League during the 1988–89 season was 6'7.16". The average weight was 216.16 pounds. There were 43 players who were 7' or taller.

Source: National Basketball Association.

Average Age in the NBA

During the 1988–89 season, the average player in the NBA was 27.01 years old.

Source: Ibid.

5

ACCIDENTS

Alcohol and Traffic Crashes

An estimated 40 percent of Americans will be involved in an alcohol-related crash at some time in their lives.

In single-vehicle fatal crashes occurring on weekend nights in 1988, 77 percent of the fatally injured drivers 25 years of age or older were intoxicated. 64 percent of drivers under the age of 25 were intoxicated.

Source: U.S. Department of Transportation, National Highway Traffic Safety Administration.

Safety-belt Usage

When safety belts are used correctly—i.e., lap and shoulder—they reduce the risk of fatal or serious occupant injury by between 40 and 50 percent.

States with mandatory seat-belt laws have 7-percent fewer fatalities than states without belt laws.

Among people over 4 years old, safety belts saved about 4,500 lives in 1988—3,800 in states with belt-use laws.

If all front-seat occupants wore safety belts in 1988, about 15,000 lives would have been saved.

Source: National Center for Statistics and Analysis.

Fatal Crashes

With about 42,119 fatal crashes in the nation in 1988, 24,711 crashes involved only 1 vehicle, and 17,408 involved 2 or more vehicles.

With a total of about 62,686 vehicles involved in fatal crashes in 1988, 36,944 were passenger cars and 13,585 were light trucks.

With a total of 47,093 persons killed in fatal crashes in 1988, 27,260 were drivers, 11,805 passengers, and 8,028 nonoccupants.

Males outnumber females as fatal-crash victims by an average of 2 to 1.

Almost half of all fatalities occur on weekends.

Source: National Highway Traffic Safety Administration's Fatal Accident Reporting System.

Accidental Death

Every 10 minutes, 2 people die in car crashes and 170 are injured.

Source: National Safety Council Estimates.

Pedestrian Deaths

In 1988, 6,869 pedestrians were killed in motor-vehicle crashes.

About 70 percent of pedestrians who were killed were males.

Source: Insurance Institute for Highway Safety.

Motorcycle Crashes

3,486 motorcyclists died in crashes in 1988, 73 percent among males age 16 to 34.

55 percent of motorcycle deaths occurred on weekends.

2 out of 5 fatally injured motorcycle drivers either

didn't have a valid license to operate their cycles or had had their licenses suspended or revoked.

Source: Ibid.

Teenage Car-Crash Deaths

In 1988, 7,244 teenagers died from motor-vehicle-crash injuries.

Teenagers constituted 15 percent of all motor-vehicle deaths and account for only 10 percent of the population.

Source: Ibid.

Safe and Unsafe Cars

The lowest death-rate car is the Volvo 740/760 four-door series.

The car with the highest death rate is the Chevrolet Corvette.

Source: Ibid.

An Airbag May Save Your Life

There is a 7-percent chance that your airbag will inflate when it's supposed to during a given 10-year period. And when it does, the impact of the crash is not one that you would have wanted to take lightly.

Source: Ford Motor Company.

Death by Unnatural Causes

About 7 percent of the 2 million Americans who die annually are killed by murder, suicide, or accident.

Source: *Time*, August 14, 1989.

Accidental Death by Airplane

The odds were 1 in 2.2 million of being killed in an airplane crash in 1988.

Source: National Safety Council.

Accidental Death by Small Commuter-Plane Crash

While flying on a major airline is considered far safer than travel by car, in 1987 you were 56 percent more likely to die in a commuter-plane crash than you were in a motor vehicle.

Source: *Good Housekeeping*, October 1989.

Daytime Car Lights Prevent Accidents

A study made with 2,000 rental cars with automatic daytime lights, as compared with an identical fleet without them, revealed that 7-percent fewer accidents occurred with the cars that kept the lights on.

Source: Insurance Institute for Highway Safety.

65 mph vs. 55 mph

When the speed limit was increased from 55 mph to 65 mph in 1988, it had its toll on fatalities on rural interstate highways. The raised speed limit increased fatalities by an estimated 31 percent.

Source: National Highway Traffic Safety Administration.

Guns Are Kid Killers

Of the children between the ages 1 to 19 who died in 1987, 11 percent were killed by guns.

Source: National Center for Health Statistics.

Airbags—More Than a Lot of Hot Air

In 1989, with an estimated 100,000 Ford vehicles equipped with airbags on the road, there was not a single report of one that released when it wasn't supposed to.

Source: Ford Motor Company.

Drunk-Driving Facts

Traffic crashes are the greatest single cause of death for every age between the ages of 5 and 32. More than half of these fatalities are a result of alcohol-related crashes.

Each year, about 600,000 (10 percent) of all police-reported motor-vehicle crashes are alcohol-related. In 1988, 47,093 people were killed in 42,119 traffic crashes.

An estimated 50 percent of these fatalities were in alcohol-related crashes (23,352 deaths).

39 percent (16,528) of these fatal crashes involved an intoxicated driver or nonoccupant.

Almost 38 percent of all fatally injured drivers were intoxicated.

Nearly 25 percent of all drivers involved in fatal crashes were intoxicated at the time of the crash.

Source: U.S. Department of Transportation. National Highway Traffic Safety Administration.

Drunk Driving and the Youth

More than 40 percent of all 15-to-19-year-old deaths result from motor-vehicle crashes. About half of these fatalities were in alcohol-related crashes. Estimates are that 3,158 persons in this age group died in alcohol-related crashes in 1988.

In 1988, nearly 27 percent of all fatally injured 15-to-19-year-old drivers were intoxicated.

Source: Ibid.

Safety Belts and Alcohol

Safety belts are used by approximately 9 percent of intoxicated drivers involved in fatal crashes (BAC 10 percent or greater), as compared with 16 percent of impaired drivers (BAC between .01 percent and .09 percent) and 28 percent of sober drivers.

Source: Ibid.

Injuries Outnumber Deaths

For every injury death:
45 children are hospitalized.
1,300 are treated and released by emergency facilities.
2,600 are treated at home.

Source: Statewide Comprehensive Injury Prevention Program, Massachusetts Department of Health.

Cycling Deaths

In 1987, 400 child cyclists died and 37,000 were injured in collisions with motor vehicles.

Source: National Highway Traffic Safety Administration's Fatal Accident Reporting System.

Head-Injured Children

1 in 7 children suffer head injuries in bike-related accidents.

Source: The National Electronic Injury Surveillance System of the Consumer Product Safety Commission.

Suffering Cyclists

On average, a cyclist involved in a motor-vehicle collision will suffer:
1.4 days in the hospital
1.4 days in bed at home
4.3 days missed from school
23.6 days of pain and discomfort.

Source: National Highway Traffic Safety Administration.

Abundance of Head Injuries

Over 2 million head injuries occur each year, with over a half million severe enough to require hospital admission. Every 15 seconds someone receives a head injury, every 5 minutes 1 of these people will die and another will become permanently disabled.

Source: National Head Injury Foundation.

Permanent Brain Injury

Of the 2 million people who suffer head injuries each year, about 2,000 will exist in a permanent vegetative state.

Source: Ibid.

Cost of Brain Injury

A survivor of a severe brain injury typically faces 5 to 10 years of intensive services at an estimated cost of over $4 million.

Source: Ibid.

Death on 2 Wheels

11 percent of Americans killed in motor-vehicle accidents were riding motorcycles.

Source: *Statistical Abstract of the United States 1987*. Bureau of Census. U.S. Department of Commerce.

Watch Your Step

14 percent of all Americans killed in motor-vehicle accidents were pedestrians.

Source: Ibid.

Homemade Accidents

43 percent of American accidents occur in the home.

Source: Current estimates from the National Health Interview Survey–Vital and Health Statistics. Public Health Services. U.S. Department of Health and Human Services: 1987.

Monkeying Around

55 percent of all American playground injuries occur on the monkey bars.

Source: Lewis Lapham, Michael Pollan, and Eric Etheridge. *The Harper's Index*. Henry Holt: 1987.

When You Swim, Don't Drink!

69 percent of all American drowning deaths involve alcohol.

Source: "Sixth Special Report to the U.S. Congress on Alcohol and Health." From the Secretary of Health and Human Services: January 1987.

What's the Weather Report?

1 out of 4 American commercial-airline accidents is caused by weather.

38 percent of American airline accidents are caused by pilot error.

Source: *The New York Times*, March 20, 1988.

Middle-Aged Adults Most Prone

One third of all injuries are in the 25-to-44 age group; the second highest incidence of injuries, more than one fifth, occurs among persons age 15 to 24.

Source: Dorothy P. Rice, *et al. Cost of Injury in the United States. A Report to Congress*. The Johns Hopkins University: 1989.

Burned to Death

Each year approximately 6,000 Americans die in fires, and 28,000 are seriously injured.

Source: *Woman's Day*, February 16, 1988.

Disabled by Accidents

50,000 children each year are disabled as a result of an accident.

Source: Alan Doelp. *In the Blink of an Eye*. Prentice Hall Press: 1989.

Causes of Fires

Smoking materials	28.9%
Heating equipment	16.8%
Arson	13.6%
Electrical	9.6%
Child's play	8.0%

Source: National Fire Protection Association.

Unintentional Accidents

Two-thirds of deaths from injuries are considered unintentional.

Source: Rice, *et al. Cost of Injury in the United States. A Report to Congress.*

Seat Belts Save Lives

Use of seat belts reduces motor-vehicle fatalities by about 45 percent and serious injuries by about 50 percent.

Source: Buckle Up for Safety Campaign.

Smoke Detectors Save Lives

Smoke detectors cut the risk of dying in fires by half.

Source: Rice, *et al. Cost of Injury in the United States. A Report to Congress.*

Abundance of Injuries

Each day more than 170,000 men, women, and children are seriously injured; nearly 400 die as a result of their injuries.

Source: National Safety Council. *Injury Prevention: Meeting the Challenge.* National Center for Health Statistics, U.S. Government Printing Office, 1987.

Injury: The Greatest Threat

Injury is the single greatest killer of Americans from age 1 to 44.

Source: National Research Council. *Injury in America: A Continuing Health Problem.* Washington, D.C.: National Academy Press, 1985.

Men More Clumsy

The risk of injury is highest among males; they sustain 57 percent of the injuries.

Source: Rice, *et al. Cost of Injury in the United States. A Report to Congress.*

Loss of Productivity

5.1 million years of productivity output has been lost as a result of injuries—which equates to about $65 billion.

Source: Ibid.

Cost of Injuries

Motor vehicle	$49 billion
Falls	$37 billion
Firearms	$14 billion
Poisonings	$9 billion
Fires and burns	$4 billion
Drownings	$2 billion
Other causes	$42 billion

Source: Ibid.

Slow Down at Crossings

The chance of death or serious injury is 40 times greater in the event of train–motor-vehicle collision than for any other type of highway accidents.

Source: Federal Railroad Administration.

Dangerous Fun

Fireworks caused 6,270 injuries in 1989. Firecrackers caused 39 percent of the injuries, sparklers 19 percent, bottle rockets 12 percent, shells 6 percent, fountains 5 percent, ground spinners 3 percent, and missiles 1 percent. Homemade, public display, and others caused 15 percent of the injuries.

Source: Consumer Product Safety Commission.

Deaths from Accidents

Of the 96,000 deaths in 1988:

49,000 were motor-vehicle accidents
12,000 were falls
5,300 were from poisoning by solids and liquids
5,000 were drownings
5,000 were from fires, burns, and deaths associated with fires
3,600 were suffocation-ingested objects
1,400 were from firearms
1,000 were poisoning by gases and vapors
13,700 were all other types

Source: *Accident Facts 1989*.

Time Lost Because of Accidents

In 1988, 35 million days were lost due to accidents.

Source: Ibid.

School-bus Accidents

110 persons nationwide were killed during the 1987–1988 school year, including 40 pupils, 5 bus drivers, and 65 other persons.

Source: Ibid.

Deaths from Fires

Cigarettes were associated with 29 percent of the 4,470 civilian fire deaths in 1989. The next most frequent causes of fire deaths were heating equipment (13 percent) and electrical distribution (12 percent).

Source: U.S. Consumer Product Safety Commission/EPHA, from data obtained from the USFA and NFPA.

Low-Risk Driving

A low-risk driver is a 40-year-old man or woman who is sober, wears a seat belt, and has a heavy car (a full-sized model, about 3,700 lbs). A high-risk driver is identified as an 18-year old, intoxicated, unbelted male in a lightweight car (a compact, about 2,300 lbs). The odds of a low-risk driver vs. a high-risk driver dying in a car wreck are 1,000 times less likely—about as safe as flying in an airplane.

Source: *Wellness Letter*, University of California, Berkeley.

Surviving an Air Crash

Many passengers do live to tell about being aboard a commercial airliner crash. Four simple tips to increase your odds are: (1) Don't wear clothing made of synthetic fibers which can melt on your skin during a cabin fire. (2) Wear shoes with low heels. (3) Stay awake and sober during the takeoff and the landing stages which are statistically the most dangerous times. (4) Curl your body up into a tight crash position to lessen the chances of being hit by debris.

Source: Federal Aviation Association.

Airline Crash Survivors

Between 1983 and 1990, of the 181 accidents involving major U.S. airlines, only 31 resulted in fatalities. There were an estimated 18,000 people involved in these accidents and less than 1,000 were killed.

Source: "With Foresight, Fliers Can Cut Risks of Travel," David J. Jefferson and Laurie McGinley. *The Wall Street Journal*, February 11, 1991.

EDUCATION

College Degrees

19 percent of American adults have a college degree.

Source: *Public Opinion*, Summer 1986.

Women Doctors

28 percent of American medical degrees are conferred on women.

Source: *Statistical Abstract of the United States 1987*. Bureau of the Census. U.S. Department of Commerce.

High School Diplomas

74 percent of American adults have a high-school diploma.

Source: *Public Opinion*, November/December 1987.

Degreeless Small-Business Owners

76 percent of American owners of small businesses do not have a college degree.

Source: *Business Week*, February 8, 1988.

White Principals

90 percent of American high-school principals are white.

Source: *The New York Times*, January 20, 1988.

Time That Kids Watch Television

The average kindergarten student has seen more than 5,000 hours of TV, having spent more time in front of the TV than it takes to earn a bachelor's degree.

Source: U.S. Department of Education.

Illiteracy Runs in the Family

Children of parents who are functionally illiterate are twice as likely as other children to be functionally illiterate.

Source: National Assessment of Educational Progress.

A Nation of Dropouts

27 percent of all U.S. high-school students, some 750,000 each year, drop out. In Japan, the rate is 5 percent, and in the Soviet Union, the rate is 2 percent.

Source: U.S. Department of Education.

Passing the Law Bar

During the past 5 years, 311,163 people took the bar examination in the United States and 57 percent passed.

Source: National Conference of Bar Examiners.

Becoming a Plebe

1 out of 10 applicants was accepted into the Naval Academy in 1989.

Source: U.S. Naval Academy.

Harvard MBAs

Out of 6,900 applications received at the Harvard Business School, 924 were accepted and 800 applicants enrolled.

Source: *Barron's Guide to Graduate Business Schools.* 6th Edition. 1988.

Educated Immigrants

Immigrants who arrived between 1970 and 1980 were 50 percent more likely than natives to have postgraduate educations; and immigrants from Asia were two and a half times more likely.

Source: *The Wall Street Journal,* January 26, 1990.

College Enrollment

College enrollment reached an all-time high of 13.5 million students during the fall of 1989. 2-year schools saw a faster growth than 4-year schools, 5.3 percent vs. 2.3 percent.

Source: National Center for Education Statistics.

Clueless Kids

More children aged 4 to 6 think Mrs. Bush was the flag's seamstress than Betsy Ross, with 29 percent casting a vote for the president's wife and 15 percent going for Betsy Ross.

Source: Playschool, Inc.

More Coeds

Women outnumber men in the college classroom; they account for 53.5 percent of the student enrollment.

Source: *Statistical Abstract of the United States 1987.*

College Dropouts

40 percent of people who go to college don't graduate.

Source: 1989 Department of Education Report.

College Testing

The nation's average performance for the Scholastic Aptitude Test (SAT) is 428 for the verbal section and 476 on the math section. Iowa has the highest combined average with 1090, and South Carolina has the lowest with 838.

Source: The College Board, U.S. Department of Education.

College Bound

48 percent of high-school students go to college after high school.

Source: Ibid.

Shortage of School Nurses

The national average is about 1 nurse for every 1,500 students, but the National Association of School Nurses suggests a ratio of 1 nurse for every 750 students.

Source: National Association of School Nurses.

Cost of Public Schools

It cost taxpayers approximately $5,000 to put a child through public school.

Source: Fortune Special Report, December 4, 1989.

Going to a Private College

Only 6 percent of U.S. families can afford to send a child to a 4-year private college without financial assistance.

Source: American College Testing Program.

Dumb College Seniors

25 percent of college seniors didn't know that Columbus landed in the New World before 1500. Only 58 percent of the seniors knew that the Civil War was fought between 1850 and 1900. 60 percent had never heard of Reconstruction, the period that followed the Civil War.

58 percent couldn't identify Plato as the author of *The Republic*.

54 percent didn't know that *The Federalist Papers* were written to promote ratification of the U.S. Constitution.

44 percent didn't know that Herman Melville wrote *Moby Dick*.

42 percent couldn't identify the Koran as the sacred text for Islam.

Source: The Gallup Organization.

Loose Requirements

77 percent of the nation's colleges and universities don't require students to take a foreign language. 44 percent have no math requirements and 38 percent have no history requirements.

Source: National Endowment for the Humanities.

Functionally Illiterate

1 out of every 5 Americans is functionally illiterate, with reading skills below the eighth-grade level. Another one third of the adult U.S. population lack the reading, writing, and comprehensive skills necessary to perform basic tasks such as filling out a job application form, reading the label on a medicine bottle, or exercising their right to vote.

Source: Laubach Literacy International.

About the Illiterate

56 percent of the illiterate are under the age of 50. 37 percent speak a non-English language at home.

Source: Ibid.

Illiterate English-Speaking Americans

50 percent did not finish high school.
42 percent had no earnings in the previous year.
35 percent are in their twenties and thirties.

Source: Ibid.

Illiterate Americans Who Speak in a Foreign Language

82 percent were born outside the United States.
21 percent entered the United States within the previous six years.
42 percent live in neighborhoods where a foreign language is the dominant language.

Source: Ibid.

Inmates Lacking Education

60 percent of prison inmates are illiterate, and 85 percent of juvenile offenders have a reading problem.

Source: Ibid.

Unemployed with Difficulties

75 percent of unemployed adults have reading or writing problems.

Source: Jonathan Kozol. *Illiterate America*. Anchor Press, 1985.

Remedial Training

Three-quarters of the Fortune-500 companies provide some level of remedial training for their workers, costing in the neighborhood of $300 million per year and affecting 8 million workers.

Source: U.S. Department of Education.

Illiterate Army Enlistees

27 percent of army enlistees can't read training manuals written at the seventh-grade level.

Source: Laubach Literacy International.

AIDS Education

Almost 80 percent of the U.S. public schools offer AIDS education. AIDS education reaches about 86 percent of seventh graders, but trails off to only about 58 percent for seniors.

Source: National School Boards Association.

Sex Ed

Condom use is discussed at about 97 percent of urban schools compared with about 89 percent of suburban schools.

Source: Ibid.

Literacy Volunteers

Volunteers to help people learn to write are 80 percent female; 82 percent of them are whites.

Source: Literacy Volunteers of America.

Starting Salaries

The starting salary for a college graduate with a bachelor's degree is $25,256; with an MBA the starting salary is $39,840. The highest starting salary is in engineering, with an average of $33,380 for chemical engineers, $32,256 for mechanical engineers, and $32,107 for electrical engineers.

Source: *USA Today*, December 11, 1989.

Animals Used for Vet Schools

Less than 10 percent of veterinary schools' curriculum now involves use of live animals for training.

Freshman Activism

About 37 percent of college freshmen participated in organized demonstrations in the 1989–1990 school year, more than double the percentage of the turbulent 1960s.

Source: Higher Education Research Institute.

Pick Up a Book!

44 percent of American adults did not read a book in the course of a year.

Source: U.S. Department of Education.

Females Outnumber Males on College Campuses

In the fall of 1990, there were nearly 14 million college students in the United States of which 55 percent were women. That's about one million more females than males on the campuses. While women have outnumbered men in the nation's colleges since 1979, the increase in their number has been almost three times the increase for males.

Source: National Center for Education Statistics.

Four-Year Universities?

Only 1 student in 6 finishes college in 4 years. After 6 years about 40 percent have a degree and about 15 percent are still in school pursuing one.

Source: National Institute of Independent Colleges and Universities.

Principals' Salaries

The average high-school principal earns $55,722 each year; the salary for a junior-high-school principal averages $52,163, and for elementary principals, the average is $48,431.

Source: The National Association of Secondary School Principals.

Teachers' Pay

The average teacher in the United States had an average salary of about $31,178 in the 1990–1991 school year. Above average:

Far West	$35,310
Mid-Atlantic region	$34,689
New England	$33,964
Great Lakes states	$33,425

Pay is below average in:

Southwest	$26,355
Southeast	$26,883
Mountain states	$27,542
Plains states	$27,874

Source: Educational Research Service.

Kids Getting Fed Up

Kids say the biggest problems at school are:

Student discipline	23%
Violence	12%
Teasing	8%
Grades	7%
Litter	5%

Source: Waldenbooks survey of 5,000 kids.

Japanese Students Have It Easy

Japanese college students go to classes and study far less than their American counterparts. About 53 percent of Japanese students say they study 2 hours or less each day and about 37 percent said they don't study at all. On the other hand, 30 percent of Americans students say they study in libraries and at home 7 hours a day or longer; 26 percent said they study 2 hours a day or less, and only .5% said they don't study at all. Japanese students also spend less time in class or at laboratories. 36 percent of U.S. students say they spend 7 hours each day at school, while only 9.8 percent of Japanese students say they spend that much time at class or lab. Only 8 percent of U.S. students spend 2 hours or less in the classroom or lab a day; among Japanese students it was 15 percent.

Source: Gakusei Engo-Kai, publisher of a job-placement magazine in Tokyo.

Private Education vs. Public Education

There are 27,000 private schools in the United States, representing 25 percent of the nation's elementary and secondary schools. The private schools are attended by just 12 percent of all students. The average private school has 195 students, while the average public school has 508 students. In 1987, 73 percent of graduates from private schools applied to colleges as compared with 48 percent of public-school graduates.

Source: U.S. Department of Education.

Bilingual Education

Bilingual education is offered in 20 percent of the public schools as compared with 7 percent in the private schools.

Source: Ibid.

Remedial Reading Classes

While 80 percent of the public schools have remedial reading classes, only 57 percent of the private schools do.

Source: Ibid.

Handicapped Programs in the Schools

91 percent of the public schools have programs for the handicapped vs. only 19 percent of the private schools.

Source: Ibid.

Postgraduation Gap

The time between graduation from college and the awarding of a Ph.D. has lengthened by 30 percent over the past 20 years, with the average gap now ranging from about 7.4 years in the physical sciences to 16.2 years in education.

Source: National Research Council.

School Days

The average number of days in the U.S. school year is 180. This compares with 180 in Sweden and Mexico. Belgium has 160; Scotland has 200; Israel has 215; South Korea has 220; and Japan has 243.

Source: National Association of Elementary School Principals.

CRIME

Jesse Lives

1 percent of American robberies are committed in a bank.

Source: *Sourcebook of Criminal Justice Statistics—1986*. U.S. Department of Justice.

"You'll Never Take Me Alive"

3 percent of Americans who murder law enforcement officers then kill themselves.

Source: Ibid.

Crime in School

5 percent of violent crime in America occurs in school.

Source: *Report to the Nation on Crime and Justice*. U.S. Department of Justice, 1983.

Fatal Attraction

6 percent of all American men who are killed are murdered by their wives or girlfriends.

Source: *Uniform Crime Reports: Crime in the United States: 1986.* Federal Bureau of Investigation. U.S. Department of Justice.

It's a Crime

In 1989, 24.9 percent of households suffered a violent crime or property crime. This includes attempted and completed crimes. Up from 24.6 percent in 1988.

Source: Bureau of Justice Statistics.

The Thieving Thirties

12 percent of all Americans 30 to 34 are arrested during a single year.

Source: *Sourcebook of Criminal Justice Statistics—1986.*

Unknown Motives

23 percent of American murders have an unknown motive.

Source: *Statistical Abstract of the United States 1987.* Bureau of the Census. U.S. Department of Commerce.

Teenage Arsonists

One quarter of American arsonists are under 15 years of age.

Source: *Uniform Crime Reports: Crime in the United States—1986.*

Female Murder Victims

26 percent of American murder victims are female.

Source: *Sourcebook of Criminal Justice Statistics—1986.*

High-School Thieves

27 percent of American high-school seniors have shoplifted in the last year.

Source: Ibid.

Guilty but Not Imprisoned

30 percent of American murderers convicted in the U.S. District Court are not imprisoned.

Source: *Statistical Abstract of the United States 1987.*

Car Thefts

34 percent of American car thefts occur in parking lots.

Source: *Sourcebook of Criminal Justice Statistics—1986.*

Female Counterfeiters

34 percent of American counterfeiters are female.

Source: *Uniform Crime Reports: Crime in the United States—1986.*

Death Behind Bars

36 percent of American inmates who die in jail die of natural causes.

Source: *Sourcebook of Criminal Justice Statistics—1986.*

How Much Longer?

36 percent of Americans on death row have been there for over 4 years.

Source: *Statistical Abstract of the United States 1987.*

Weaponless Robbers

43 percent of all American robbers do not use weapons.

Source: *Uniform Crime Reports: Crime in the United States—1986.*

Bad Guys with Guns

One half of all American murders are committed with a handgun.

Source: *Report to the Nation on Crime and Justice.*

Nighttime Crime

54 percent of American robberies are committed during the night.

Source: *Sourcebook of Criminal Justice Statistics—1986.*

White Victims

56 percent of all American murder victims are white.

Source: Ibid.

Hang 'Em, Gas 'Em, Etc.

67 percent of all Americans favor the death penalty for convicted murderers.

Source: General Social Surveys, 1972–1987. Conducted for the National Data Program for the Social Sciences at National Opinion Research Center, University of Chicago.

Have Gun, Will Shoot

69 percent of Americans who own a firearm say that if a burglar broke into their home at night, they would use it.

Source: *Sourcebook of Criminal Justice Statistics—1986.*

Lock Your Doors

1 out of 4 U.S. households experienced a burglary or theft in 1988 or had some household member victimized by a crime of violence or theft.

Source: U.S. Department of Justice.

Doing Time

On June 30, 1988, the jail population in the United States totaled 343,569, or 144 of every 100,000 Americans. Non-Hispanic whites constituted 43 percent, non-Hispanic blacks 41 percent, and Hispanics 15 percent.

Source: Ibid.

Murder and Non-negligent Manslaughter in the United States

In 1988, there were 20,675 people killed by other people in the United States, or about 8.4 murders per 100,000 inhabitants.

Source: U.S. Department of Justice, Federal Bureau of Investigation.

Death Row

During 1988, 2,124 prisoners in the United States were in prison with death sentences. 11 prisoners were executed in 6 states—Florida, Georgia, Louisiana, Texas, Utah, and Virginia. The median time since the death sentence was imposed for the 2,124 prisoners was 3 years and 9 months. 58.3 percent were white; 40.2 percent were black; 1.0 percent were American Indian; and .6 percent were Asian. 1.1 percent were female. The median age of all inmates under a death sentence was nearly 33 years.

Source: U.S. Department of Justice.

Escapees from Prison

In 1988, some 10,351 convicts escaped from U.S. prisons and only 76 percent were recaptured.

Source: *Corrections Digest.*

Drug Traffickers Have It Easy

The average stay of 22 months for a drug trafficker is less than the average sentence for those convicted of aggravated assault or robbery.

Source: National Institute on Drug Abuse.

Violent Crime in the United States

In 1989, a total of 4.9 percent of households experienced a violent crime with 3.9 percent suffering assault, 1 percent robbery, and 0.1 percent rape. Up from 4.8 percent in 1988.

Source. U.S. Department of Justice.

Sentencing by State Courts

State courts send about 46 percent of convicted felons to state prisons, 21 percent to local jails, sentence 31 percent to straight probation, and give 2 percent other nonincarceration sentences.

Source: National Judicial Reporting Program of the U.S. Department of Justice.

Sentencing of Drug Traffickers

Nationwide, state courts send 37 percent of drug traffickers to prison, 27 percent to local jails, and sentence 35 percent to straight probation.

Source: Ibid.

Sentenced to Death

About 2 percent of those convicted of murder are sentenced to death.

Ibid.

Time in the Slammer

The average time spent incarcerated is 4 years, 10 months; the median sentence is about 3 years. State prisons have an average stay of 6 years, 9 months, while local jails have an average stay of 9 months.

Source: Ibid.

Doing Time Again

Of the prisoners released from jail, 63 percent of them return because of repeated offenses. About 47 percent were reconvicted and 41 percent returned to jail.

Source: U.S. Department of Justice.

Women Prisoners

29,000 women spent time in jail in 1988. 8 in 10 of them were mothers, most of them unmarried.

Source: Ibid.

Youth Incarceration

In 1987, 53,503 youths were held in public juvenile facilities. 86 percent were male and 82 percent were between the ages of 14 and 17. Almost 43 percent of the youths in custody had been arrested more than 5 times.

Source: Ibid.

White-Collar Crimes

When people are tried for a white-collar crime, nearly 3 out of 4 are convicted, but studies suggest that only about 15 percent of such crimes are ever detected.

Source: The United Way.

Crime Under the Influence

35 percent of prison inmates said that they were under the influence of drugs when they committed their offense. Nearly 40 percent of youths were under the influence of drugs when they committed their offense.

Source: Bureau of Justice Statistics.

Time for Convicted Felons

Type of Crime	Average Time Sentenced
Murder and non-negligent manslaughter	7 years, 2 months
Rape	5 years, 6 months
Robbery	4 years, 9 months
Burglary	2 years, 7 months
Drug trafficking	1 year, 10 months

Source: The National Judicial Reporting Program of the U.S. Department of Justice.

Making It to the Federal Courts

About 5 percent of felony convictions go to the federal courts.

Source: *Drug Law Violators, 1980–1986: Federal Offenses and Offenders.* Bureau of Justice Statistics Special Report. June 1988.

Homicide Rates

In 1988, 8.4 people were killed per 100,000 population.

Source: Canadian Centre for Justice Statistics, Montreal Police.

DNA Fingerprinting

The odds of 2 people having the same DNA, other than identical twins, is 30 billion to 1.

Source: *Columbus Dispatch*, June 18, 1989.

Generation Rapes

21 percent of women who are raped as youngsters are again raped as adults.

Source: John Briere of the University of Southern California, and Jon Conte of the University of Chicago.

Getting Battered Again

17 percent of women who are battered as youngsters are again battered as adults.

Source: Ibid.

Don't Kill Me, Cousin John

58 percent of known murderers were relatives or acquaintances of the victim.

Source: *Report to the Nation on Crime and Justice*.

Don't Walk Alone at Night

73 percent of rapes occur at night between 6:00 P.M. and 6:00 A.M.

Source: Ibid.

Why Murder?

39 percent of murders are a result of arguments; 18 percent are a result of a felony; and 9 percent are a result of a robbery.

Source: Ibid.

Lock Your Doors

The average theft loss from robbery was $447, and guns were actually discharged in a fifth of all robberies.

Source: *Robbery Victims*. Bureau of Justice Statistics Special Report. April 1987.

Lock Your Doors—at Least in August and December

Robberies are most likely to occur in August and December and least likely to occur in February and April.

Source: *Report to the Nation on Crime and Justice*.

Who Did It?

A household member is present in about 10 percent of all burglaries.

Source: *National Crime Survey*. Bureau of Justice Statistics.

Crimes in the Home

Persons who illegally enter homes commit:
three-fifths of all rapes in the home
three-fifths of all robberies in the home
about a third of all aggravated and simple assaults in the home.

Source: Ibid.

Automatic-Teller-Machine Fraud

Lost or stolen cards are used in about 50 percent of all automated-teller-machine frauds.

Source: *Electronic Fund Transfer Fraud*. Bureau of Justice Statistics Special Report. March 1985.

Where's My Car?

About 1 percent of all households reported the theft or attempted theft of a car in 1985.

Source: *National Crime Survey*. Bureau of Justice Statistics.

Pocket Picking

Personal larcenies such as pocket picking occur on the street 23 percent of the time, in commercial buildings 22 percent, and on public transportation 15 percent of the time.

Source: *Criminal Victimization in the United States: 1985.* Bureau of Justice Statistics Special Report. May 1987.

Dangerous Parking Lots

36 percent of violent crimes such as rape, robbery, and assault, and 28 percent of personal larceny such as pick pocketing happen on a street or parking lot.

Source: Ibid.

Weapons in Crimes

Except for homicide, most violent crimes do not involve weapons.

Weapon Used	Homicide	Rape	Robbery	Assault
A weapon used	93%	25%	49%	34%
Firearm	59%	11%	23%	12%
Knife	21%	14%	21%	10%
Other	13%	5%	13%	15%
None used	7%	75%	51%	66%

Source: *The Use of Weapons in Committing Crimes.* Bureau of Justice Statistics Special Report. January 1986; and FBI Crime in the United States: 1985.

Assaults to Police Officers

Means of Assault	Percentage of All Assaults
Firearm	5%
Knife	3%
Other weapon	9%
Hands, feet, fist, etc.	84%

Source: FBI Law Enforcement Officers Killed and Assaulted, 1985.

Victims of Crime

Of rape, robbery and assault victims:
30% were injured
15% required some sort of medical attention
8% required hospital care

Source: *Report to the Nation on Crime and Justice.*

Who Are the Victims?

Violent crime is more likely to strike:
Men than women except for rape
The young than the elderly
People with low rather than high incomes
Blacks than whites or members of other minority
groups
The divorced or separated and the never married than
the married or widowed

Source: *Criminal Victimization in the United States.*

Risk of Being a Homicide Victim

Lifetime risks of being a homicide victim:
1 out of:
179 for males
30 for black males
495 for white females
132 for black females

Source: Updated data based on similar material from *The Risk of Violent Crime*. Bureau of Justice Statistics Special Report. May 1985.

Odds of Being a Victim of a Violent Crime

An estimated five-sixths of us will be victims of attempted or completed violent crimes during our lives. The risk is greater for males than females and for blacks than whites.

Source: *Lifetime Likelihood of Victimization*. Bureau of Justice Statistics Technical Report. March 1987.

Household Burglaries

The odds of a household being burglarized over a 20-year period are 72 percent, larceny 89 percent, and a motor-vehicle theft 19 percent.

Source: Ibid.

Preventing Crime

25 percent of people have engraved valuables to facilitate identification.

7 percent have participated in a neighborhood watch program.

7 percent have installed a burglar alarm system.

Source: The Victimization Risk Survey.

Self-Defense

Victim's Response	Rape	Robbery	Assault
Weapon used	3%	4%	4%
Physical force	29%	22%	23%
Verbal response	19%	10%	16%
Attracting attention	18%	13%	10%
Nonviolent evasion	13%	14%	23%
Other	1%	2%	3%
No self-protective action	18%	37%	22%

Source: *National Crime Survey, 1979–1985*. Bureau of Justice Statistics.

Unreported Crime

More than two-thirds of crime victims do not report their crime to police.

Source: Bureau of Justice Statistics.

Abundance of Child Abuse

More than 2.3 million children are victims of child abuse and neglect.

1,225 kids died in 1988 as victims of child abuse and neglect.

Source: The American Association for Protecting Children.

Elderly Abuse

An estimated 1.1 million elderly Americans are abused annually, and of these nearly 86 percent are abused by members of their own family.

Source: The United Way.

Spouse Abuse

Between 2 million and 6 million women are battered each year and of that 3 in 4 victims of spouse abuse were divorced or separated at the time of the abuse.

Source: National Coalition Against Domestic Violence.

Hereditary Abuse

Of the boys and girls raised in abusive households, an estimated 60 percent of boys will grow up to be batterers, and 50 percent of the girls will grow up to be battered women.

Source: Ibid.

Neighborhood Watch Guards

There are approximately 18,000 neighborhood-crime-watch groups in the United States, and evidently they have a high impact. Police estimate that crime rates drop an average of 35 percent in these areas.

Source: National Association of Town Watch.

Neighborhood Watch Programs

1 in 5 families lives in an area with a neighborhood watch program.

38 percent of the households in these areas participate in the program.

Source: *Report to the Nation on Crime and Justice.*

Reporting to the Police

Women who get their purses snatched, losing over $250 or more, report it to the police 70 percent of the time compared with only about 55 percent who report completed rape.

Source: *Reporting Crimes to the Police.* Bureau of Justice Statistics Special Report. December 1985.

Victim-Offender Relationship

Relationship	Homicide	Robbery	Assault
Stranger	18%	75%	51%
Acquaintance	39%	17%	35%
Relative	18%	4%	10%
Unknown	26%	4%	4%

Of the violent crimes against strangers, 70 percent were against males, and of the violent crimes against relatives, 72 percent were against females.

Source: *Report to the Nation on Crime and Justice.*

Corrupt Citizens

16 to 18 percent of the population have arrest records for nontraffic violations.

Source: Ibid.

Released Rapists

Released rapists are more than 10 times more likely than nonrapists to be rearrested for rape.

Source: U.S. Department of Justice.

Released Murderers

Released murderers are 5 times more likely than other offenders to be rearrested for homicide.

Source: Ibid.

Youngsters Going Back for More

More than 75 percent of those 17 and younger who are released from prison are rearrested.

Source: Ibid.

Average Age of Arrest

Charge	Average Age (1985)
Gambling	37 years
Murder	30 years
Sex offenses	30 years
Fraud	30 years
Embezzlement	29 years
Aggravated assault	29 years
Forcible rape	28 years
Weapons	28 years
Forgery	27 years
Drug-abuse violations	26 years
Stolen property	25 years
Theft	25 years
Arson	24 years
Robbery	24 years

Charge	Average Age (1985)
Burglary	22 years
Motor-vehicle theft	22 years

Source: "Age-Specific Arrest Rates and Race-Specific Arrest Rates for Selected Offenses 1965–1985." FBI Uniform Crime Reporting Program. December 1986.

Carry Mace

Nearly 1 out of 12 females will be the victim of a completed or attempted rape sometime during their lives.

Source: U.S. Department of Justice.

Being Examined

Almost 3 percent of the adult male population is on probation or parole.

Source: Ibid.

Waiting to Die

The average length on death row is 6 years and 8 months for those who are executed. The median time for all prisoners on death row is 3 years and 9 months.

Source: Ibid.

Cop Killings

The odds of a law enforcement officer in the United States being killed in the line of duty is 3,797 to 1 in any given calendar year.

Source: *Police Times*.

Death-Row Inmates

Of the 2,186 death-row inmates in 1989, almost 99 percent are male.

Source: American Correctional Association.

Drugs and Crime

75 percent of jail inmates, 79.5 percent of state prisoners, and 82.7 percent of youth in long-term juvenile facilities had used drugs at some point in their lives.

Source: U.S. Department of Justice, Bureau of Justice Statistics.

Serving More Than a Year

The chance of being sentenced for more than a year in prison for those arrested for:
Homicide was 49%
Rape was 29%
Robbery was 28%
Burglary was 20%
Motor-vehicle theft was 10%
A felony weapons violation was 7%.

Source: Ibid.

1 out of 5 Released Felons Repeat Offense Before Trial

Nearly 1 out of 5 people arrested and charged with committing a felony are released from custody and rearrested for a similar felony offense before going to trial. Of those rearrested on additional felony charges, about two thirds were again released into the community!

Source: U.S. Department of Justice.

Black Male Lawbreakers

Nearly 1 out of 4 black males in his twenties is behind bars, on probation, or on parole.

Source: The Sentencing Project.

Inmates Were Abused Children

4 out of 5 prison inmates were abused as children.

Source: National Center for the Prevention of Child Abuse and Neglect.

Abusive Parents

An estimated 5 percent of parents in the general population abuse their children, while about 97 percent of parents spank or hit their children.

Source: Edward Zigler, a Yale psychology professor, and graduate student Joan Kaufman.

Specialized Car Theft

New York has 10 percent of the nation's Toyota Cressidas but 58 percent of the theft claims for the vehicle. Texas has only 7 percent of the nation's Grand Marquis but 80 percent of the theft claims for the vehicle.

Source: Highway Loss Data Institute.

Sexual Abuse of the Young

Female children are abused 4 times as often as male children.

Source: Study of National Incidence and Prevalence of Child Abuse and Neglect, 1988.

Family Income and Abuse

Children from families with incomes of less than $15,000 had an overall rate of maltreatment 5 times that of other children: 32.3 vs. 6.1 children per 1,000. Neglect is nearly 8 times higher for children of low-income families.

Source: Study of National Incidence and Prevalence of Child Abuse and Neglect: 1988.

Dealing in Drugs Is Dangerous

The odds of getting killed are 1 in 70 for people who sell drugs in the United States. The odds are 1 in 14 of becoming seriously disabled, and there's a 22-percent chance of going to jail.

Source: The Rand Corporation.

Seizing Cocaine

The Drug Enforcement Agency seized 56,980 kilograms of cocaine in 1988, 50 percent more than in 1987.

Source: *Statistical Abstract of the United States 1989*. Bureau of the Census. U.S. Department of Commerce.

Dishonest Workers

Employee theft is such a problem now that workers are stealing more than shoplifters. Nearly 1 in every 15 retail workers is caught stealing, and retailers only nab an estimated 25 percent of those who do steal.

Source: Jack L. Hayes International.

Women Guards

21 percent of American guards are women.

Source: *Statistical Abstract of the United States 1987*.

S&L Scams

The average savings-and-loan offender in 1989 was sentenced to 1.9 years, while the average bank robber was sentenced to 9.4 years.

Source: Federal Bureau of Investigation.

Theft in the United States

17.8 percent of households suffered theft, including 11.2 percent personal theft and 8.1 percent household theft; 5.1 percent suffered a burglary; and 1.6 percent motor-vehicle theft.

Source: Bureau of Justice Statistics.

Black and White Victims

In 1989, a total of 29.2 percent of all black households suffered a surveyed crime—up from 26 percent in 1988. And .4 percent of them suffered a violent crime, a decrease from 6.6 percent in 1988.

Households headed by Caucasians showed 24.3 percent suffered a surveyed crime, up from 23.9 the previous year. And 4.7 percent were victimized by a violent crime, a slight drop from 4.6 percent in 1988.

Source: Ibid.

What Are the Odds That a Lie Detector Test Will Be Accurate?

When a lie detector test is given by a skilled individual who has attended a recognized polygraph school, and it is conducted as he was trained, the results will be accurate about 98 percent of the time. When a 10-point fingerprint is taken at the scene of the crime, the rate of accuracy is 100 percent. Studies have shown that eyewitnesses at the scene of the crime give accurate testimony only 52 percent of the time.

Source: U.S. Department of Defense.

Handgun Killings

The United States leads the world in handgun related killings. According to the latest figures that were released in 1989, the United States was well ahead of other "civilized" nations:

Country	Killings
Australia	13
Canada	8
Great Britain	7
Israel	25

Country	Killings
Sweden	19
Switzerland	53
United States	8,915

Source: Handgun Control, Inc.

The High Cost of Shoplifting

What shoplifters take, we all pay for. In 1989, the average worth of goods was $196 per shoplifter caught in the U.S. Shoplifting losses totaled $2.2 billion. 55 percent of the retailers seek civil restitution; that's a total of 76,000 cases filed each year. Each U.S. family pays more than $200 in added costs as a result of shoplifting.

Source: Ernst and Young; G.S. Schwartz and Co.

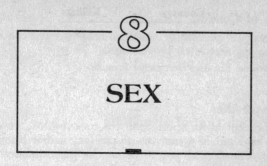

SEX

The Average Age for First-Time Intercourse in the United States

The average age of first intercourse is 17.2 for females and 16.5 for males. By age 20, 3 out of 4 girls and 5 out of 6 boys will have had sexual intercourse at least once, as will 1 in 6 15-year-old boys and 1 in 20 15-year-old girls.

Source: Ann Landers, Universal Press Syndicate.

Teen Pregnancies

The United States has twice the number of teen pregnancies as other industrialized nations. More than 80 percent of teen pregnancies are unintended.

Source: Jill Clayburgh. *Ask Me Anything: How to Talk to Kids About Sex.* Video.

Rate of Pregnancies with No Birth Control

60 to 80 percent of women who use no method of birth control will have a pregnancy within 1 year.

Source: Planned Parenthood Federation of America, Inc.

Rate of Condom Failures

Nearly 14 percent of American couples using condoms are having accidental pregnancies.

Source: The Alan Guttmacher Institute.

Transvestites

1 percent of American men are transvestites. An estimate of 1 out of 4 transvestites are gay and/or bisexual. Closet transvestites are 90 percent of the estimated 1.3 million cross-dressers in the United States. An estimated 20 percent of the transvestites are married.

Source: North American Transvestite-Transsexual Society.

The Gay Population in America

An estimated 37 percent of all (white) males have had a homosexual experience to orgasm, but only 4 percent of the men surveyed were exclusively homosexual throughout their lives. And 8 percent were exclusively homosexual for at least 3 years between the ages 18 and 65.

Source: National Gay and Lesbian Task Force (quoting the 1948 *Kinsey Report*, confirmed by a 1988 report by the National Association of Science, based on a 1970 survey).

"It's Okay . . . As Long As You Love Me. . . ."

53 percent of Americans think premarital sex between a teenage boy and a girl he doesn't love is always wrong, while 68 percent think premarital sex between a girl and boy she doesn't love is always wrong. The figures are 37 percent if the boy is in love with the girl and 46 percent if the girl loves the boy.

Source: Albert Klassen. *Sex and Morality in the United States.* Wesleyan University Press, 1989.

"Safe Days"

3 out of 5 adult women don't know when they are most likely to become pregnant, and the figure is 4 out of 5 for teenage girls. By the way, pregnancy is most likely to occur when intercourse takes place 14 days after the first day of menstruation.

Source: "Adolescent Pregnancy and Parenting in California." University of California at San Francisco.

Hitting Puberty

Adolescent girls hit puberty at about 12 years old compared with about 14 for boys.

Source: *Masters and Johnson on Sex and Human Loving*. Little Brown, 1988.

Precocious Puberty

The odds are about 1 in 10,000 that pubertal changes will occur before the age of 9.

Source: Ibid.

Classified Love

Ads that women place in the newspaper get an average of 49 responses, while these same fun-times ads placed by men get only about 15 replies.

Source: Ibid.

Sex After Sixty

More than 65 percent of couples over 60 years old and 75 percent of couples over 70 have very little or no sex.

Source: Ann Landers. *Columbus Dispatch, June 25, 1990.*

Reaching Menopause

The average women hits menopause anywhere between the ages of 48 to 52.

Source: *Masters and Johnson on Sex and Human Loving.*

Masturbation

92 percent of males masturbate and about 62 percent of females have masturbated at least once in their lives.

Source: *The Kinsey Report.*

Nocturnal Emissions

83 percent of all males and about 37 percent of females have had a "wet dream" during their sleep by the age of 45.

Source: *The Kinsey Report.*

Premarital Sex

79.4 percent of males have had at least 1 sexual experience before marriage and 40.7 percent of females.

Source: Albert Klassen. *Sex and Morality in the United States.* Wesleyan University Press, 1989.

Women Who Have Orgasms

About 88 percent of women have orgasms.

Source: *The Hite Report.*

Climaxing from Intercourse

Approximately 30 percent of women reach orgasm by means of intercourse.

Source: Ibid.

Tying the Tubes

Complications may develop in 1 to 4 percent of operations performed through the abdomen, and in 2 to 13 percent of operations performed through the vagina.

Source: Planned Parenthood.

Untying the Tubes

For a woman who regrets being sterilized and gets a reversal, the success rate for having a pregnancy is about 60 percent when the operation is performed by highly qualified experts.

Source: Ibid.

Ending Pregnancies

In the United States, women choose to end approximately 30 percent of their pregnancies. Most women (90% percent) obtain abortions in the first 12 weeks (trimester) of pregnancy. One half obtain abortions in the first 8 weeks.

Source: National Abortion Federation.

Teens Take Risks

Only 1 out of 3 sexually active teens uses birth control regularly.

Source: Jill Clayburgh. *Ask Me Anything: How to Talk to Kids About Sex*. Video.

Preventing Pregnancies

94 percent of today's American women have used some sort of birth control.

Source: Louise Tyrer, M.D., Vice-President for Medical Affairs, Planned Parenthood.

Regretting Sex

Only 16 percent of all men regret having premarital sex while 51 percent of women under 35 regret it. 39 percent of women over 35 regret it.

Source: *Psychology Today*.

Happily Married

The "very happily" married have sex 75 times a year; the not-too-happy only about 43 times per year.

Source: National Opinion Research Center, University of Chicago.

Sex Ed

40 percent of Americans first learn about sex from their friends.

Source: Louis Harris. *Inside America*. Vintage, 1987.

Sex or Money

47 percent of all American men enjoy sex more than money.

Source: Lewis Lapham, Michael Pollan, and Eric Etheridge. *The Haper's Index*. Henry Holt, 1987.

Effectiveness of Birth-Control Method

	Number of Pregnancies per 100 Women During One Year of Use
Voluntary Sterilization	Less than one
The Pill	2.5
IUD	4
Condom with foam	10
without foam	
unmarried women	11
married women	14

	Number of Pregnancies per 100 Women During One Year of Use
Diaphragm/Cervical Cap	18
Contraceptive Sponge	18
Foams, Creams, Jellies, and Suppositories	20
Withdrawal	20
Fertility Awareness Methods	24
No Method	60–80

Source: The American College of Obstetricians and Gynecologists.

Women Who Abort

70 percent of the women who have an abortion are white. 31 percent of abortions in 1987 were women attending school; 68 percent were employed women; 33 percent had family incomes of less than $11,000.

Source: National Abortion Federation.

Getting Around

The average adult has 1.16 sexual partners a year.

Source: National Opinion Research Center, University of Chicago.

Getting Nothing

22 percent of American adults have been celibate over the past year.

Source: Ibid.

Extramarital Sex

An estimated 1.5 percent of married people have had an affair in the past year. Since marriage, 1 out of 3 women and 2 out of 3 men have been unfaithful.

Source: Ibid.

Lesbian Sex

2 out of 3 homosexual women have orgasms in 90 to 100 percent of their lesbian contacts compared with only 2 out of 5 women in their fifth year of marital intercourse.

Source: *Masters and Johnson on Sex and Human Loving*.

Zoophilia

8 percent of adult men and 3.6 percent of adult females have had sexual contact with animals; for females this is usually a household pet and for males it is a farm animal.

Source: Ibid.

S&M

5 to 10 percent of men and women get sexual pleasure out of sadism and masochism.

Source: Ibid.

Odds of Impotency

The Odds (Against unless noted)

Under age 30	124 to 1
By age 40	51.6 to 1
By age 50	13.9 to 1
By age 60	4.4 to 1
By age 65	3 to 1
By age 70	2.7 to 1
By age 75	1.2 to 1 for
By age 80	3 to 1 for

Source: *The Kinsey Report*.

Dirty Pictures

About 75 percent of people use pornography to stimulate themselves sexually.

Source: Lillian Rubin. *Erotic Wars: What Happened to the Sexual Revolution*. Farrar, Straus & Giroux, 1990.

Open Marriages

5% of American couples engage in open marriages.

Source: Ibid.

Curious About Sex with the Same Sex

10 percent of Americans have experimented with sex with a member of the same sex, although all consider themselves heterosexual.

Source: Ibid.

Bondage

25 percent of people have experimented with bondage.

Source: Ibid.

Oral Sex in America

90 percent of American couples engage in oral sex.

Source: Ibid.

Women and Premarital Sex

A 1988 government survey of women 15 to 44 indicate 72.4 percent of them began their sexual experience in nonmarital relations.

16 percent say they were married at the time they first experienced intercourse, while 12 percent hadn't yet had sexual experience.

Source: National Center for Health Statistics.

Monogamous Americans

A survey shows the vast majority of Americans are monogamous, in the sense of having had only 1 sexual partner, married or unmarried, during the preceding year. Only 18 percent of all sexually active adults in a large 1988 sampling weren't monogamous according to this definition.

Almost half of the surveyed unmarried men under 30 specified only 1 sex partner during the previous 12 months.

Source: National Opinion Research Center, University of Chicago.

Sex Changes

51 percent of the sex changes in the United States are women who become men. Of course, 49 percent are men who are changed into women. Since 1965, between 40,000 and 50,000 people have had sex changes.

Source: North American Transvestite-Transsexual Society.

Women Who Use Contraception

About 60 percent of the 57.9 million U.S. women who were 15 to 44 years of age in 1988 practiced contraception; 24 percent of them relied on either female sterilization or vasectomy, while 37 percent used reversible methods.

Source: National Center for Health Statistics.

Women Orgasms

Approximately 10 percent of all women have never had an orgasm. Between 50 to 75 percent of women who have orgasms by other types of stimulation do not have orgasms when the only form of stimulation is penile thrusting during intercourse.

Source: Ibid.

Women Who Perform Fellatio

An estimated 50 to 80 percent of women perform fellatio; however, only 35 to 65 percent find it pleasurable. The rest are indifferent or do not enjoy it.

Source: Ibid.

"Withdrawal" Is Not a Reliable Source of Birth Control

About 25 percent of the women who rely on withdrawal as a form of contraception for 1 year become pregnant.

Source: Kinsey Institute for Sex, Gender and Reproduction.

It's a Boy!

The odds of giving birth to a boy are slightly less than 50 percent.

Source: Ibid.

Playing It Safe

65 percent of all teen females were using contraception, mainly condoms, during first intercourse.

Source: *Family Planning Perspectives,* Alan Guttmacher Institute. September/October 1990.

Teenage Sex (Girls)

52 percent of white, non-Hispanic teen females are sexually active. 61 percent of the black females are, and 49 percent of the Hispanic teen females are.

Source: Ibid.

Multiple Sex Partners

58 percent of all females between ages 15 through 19 have had sex with 2 or more people. 71 to 72 percent of the women between ages 20 through 34 have, as well as 57 percent of the women between 40 to 44.

Source: Ibid.

HEALTH

Color-Blindness

1 out of 100 Americans is color-blind.

Source: Current Estimates from the National Health Interview Survey. Vital and Health Statistics. Public Health Service. U.S. Dept. of Health and Human Services: 1987.

Ulcers

1 out of 50 Americans has an ulcer.

Source: Ibid.

Asthma

1 out of 25 Americans has asthma.

Source: Ibid.

Doc Shy

1 out of 25 Americans hasn't seen a doctor in at least 5 years.

Source: Ibid.

Heart Conditions

2 out of 25 Americans have heart conditions.

Source: *Statistical Abstract of the United States 1987*. Bureau of the Census. U.S. Department of Commerce.

Heart Murmurs

6 out of 25 Americans with heart murmurs are under 18 years of age.

Source: Current Estimates from the National Health Inverview Survey.

Heart-Disease Casualties

38 percent of Americans die of heart disease.

Source: *Statistical Abstract of the United States 1987*.

AIDS Victims

1 out of 4 American AIDS victims is an intravenous drug user.

Source: *Health United States 1986*. Public Health Service. U.S. Department of Health and Human Services.

Four Eyes

47 percent of Americans wear eyeglasses.

Source: *Statistical Abstract of the United States 1987*.

"Who Me? I'd Never Get AIDS."

58 percent of Americans are not concerned about getting AIDS.

Source: *Society*, January/Febraury 1988.

Weight Watchers

68 percent of American grocery stores have a diet section.

Source: *Progressive Grocer*, April 1987.

Abortions on Unmarried Women

81 percent of American abortions are performed on unmarried women. 63 percent have never been married and 18 percent are divorced.

Source: *Statistical Abstract of the United States 1987*.

America's Biggest Pain

An estimated 37 million Americans have arthritis. It's our nation's number-1 crippling disease, and your chance of getting it is 1 in 7.

Source: Arthritis Foundation.

Quack Remedies

1 out of 10 people who try "quack remedies" suffer harmful side effects. The top-10 health frauds are:

Fraudulent arthritis products
Spurious cancer clinics
Bogus AIDS cures
Instant weight-loss schemes
Fraudulent sexual aids
Baldness remedies
Nutritional schemes
Chelation (arterial cleaning) therapy
Unproven use of muscle stimulators
Needless treatments for nonexistent yeast infections

Source: U.S. Food and Drug Administration.

Herpes Prevails

An estimated 26 to 31 million Americans are infected with genital herpes.

Source: *Good Housekeeping*, October 1989.

Odds of Conception

Even under the best of circumstances, the chances of a couple conceiving during any given month is about 1 in 5.

Source: *The Kinsey Institute New Report on Sex*.

Senior Citizens Without Teeth

42 percent of Americans over age 65 have lost all of their teeth.

Source: American Dental Association.

Senior Citizens with Teeth Missing

The average American over age 65 has lost 10 teeth.

Source: American Dental Association.

Infant Mortality in the United States

The national rate of infant mortality (between birth and age 1) is 10.1 deaths per 1,000 live births. In 1987, the rate was 8.6 deaths per 1,000 for white babies and 17.9 deaths per 1,000 for black babies.

Source: National Commission to Prevent Infant Mortality.

Cocaine Birth Defects

The odds are 5 times greater that women who use cocaine early in their pregnancies will have children with urinary-tract defects. The defects run about 7.2 per 1,000 births for infants born to such mothers.

Source: Centers for Disease Control.

Common Allergies to Foods

The most common things we're allergic to that we eat are:

Dairy	40%
Seafood	21%
Vegetables	20%
Fruits	20%
Chocolate	11%

Source: NPD/Home Testing Institute.

Pregnancy Risks According to Age

Maternal Age	Risk of Down's Syndrome	Total Risk for Chromosome Abnormalities
20	1 in 1,667	1 in 526
22	1 in 1,429	1 in 500
24	1 in 1,250	1 in 476
26	1 in 1,176	1 in 476
28	1 in 1,053	1 in 435
30	1 in 952	1 in 384
32	1 in 796	1 in 322
34	1 in 500	1 in 260
36	1 in 294	1 in 164
38	1 in 175	1 in 103
40	1 in 106	1 in 65
42	1 in 64	1 in 40
44	1 in 38	1 in 25

Source: *American Journal of Obstetrics and Gynecology*, 1989.

Older Moms

7 percent of the pregnancies are among women older than 35 at delivery.

Source: Mark Landon of the Ohio State University, Columbus.

Spontaneous Abortions Increase with Age

The odds of spontaneous abortions rise with age. The rate is 12 to 15 percent for a woman in her early twenties as compared to 25 percent for a woman who is 40.

Source: Dr. Joe Leigh Simpson, Chairman of Obstetrics and Gynecology, University of Tennessee School of Medicine, Memphis.

Good Healthy Intentions

Nearly 3 out of 4 of the people who own home exercise equipment don't use it as much as they intended.

Source: The Roper Organization.

Death by Heart Disease

Heart disease claims nearly 1 million lives each year in America.

Source: The United Way.

Health-Care Costs

The United States spends almost $1.5 billion a day on health care. In 1987, the United States spent $500 billion on health care, nearly 12 percent of the U.S. GNP.

Source: National Leadership Commission on Health Care: 1989.

The Cost of Medical/Dental Insurance

The average cost of medical/dental plans in the United States was $2,748 in 1989.

Source: A Foster Higgins and Company Survey of 1,943 employees.

Jerry's Kids

Over 37 million Americans over 15 years old suffer some functional physical limitation, and 13.5 percent more are severely limited.

Source: The United Way.

Where's the Nurse?

In 1990, there is a shortage of an estimated 390,000 nurses.

Source: U.S. Bureau of Labor Statistics, 1989.

Long-Term Care

75 percent of long-term care of the elderly takes place in the home.

Source: Population Reference Bureau; 1985.

Alcohol Abuse

1 of 3 families currently reports alcohol abuse by a family member. More than half of all alcoholics have an alcoholic parent. In up to 90 percent of child-abuse cases, alcohol is a significant factor.

Source: National Association for Children of Alcoholics.

Sneezing and Wheezing

1 in 6 people is affected by asthma and allergy diseases. 1 in 5 children going to a pediatrician has a major allergic disorder. If 1 parent is allergic, the child has a 25-percent chance of developing an allergy; if both parents are allergic, almost 2 out of 3 children will be allergic also.

Source: The National Institute of Allergy and Infectious Disease.

Keeping Alcoholism in the Family

Children of alcoholics have a 4 times greater risk of developing alcoholism than children of nonalcoholics. The average age for kids' first drinking experience is 12 years old.

Source: National Council on Alcoholism.

Women Drunks

60 percent of women age 18 and older drink. Among blacks that drink, twice as many black women vs. men have health problems due to drinking.

Source: Ibid.

Birth Defects/Alcohol

Fetal alcohol syndrome is the leading known cause of birth defects that can be prevented.

Source: Ibid.

On-the-Job Injuries

The rate of injuries and illnesses for full-time workers was 8.6 per 100 in 1988. The highest rate of injuries came in the automobile industry, where 19.5 injuries were reported for every 100 full-time workers in 1989.

Source: U.S. Bureau of Labor Statistics.

X Rays in Infancy Cause Cancer

Exposure to medical X rays during infancy increases the chance of women getting breast cancer when they reach their thirties by more than 4 times.

Source: Dr. Nancy G. Hildreth of the University of Rochester School of Medicine and Dentistry.

Losing It

Over half of nursing-home patients in the United States suffer from dementia to some extent. 70 percent are probably due to Alzheimer's and/or multi-infarct dementia.

Source: Mortimer Dolman, M.D., Heritage House Medicine Directory.

Alzheimer's

47 percent of people over age 85 may have Alzheimer's; 1 in 10 people over age 65 suffer from the disease.

Source: Dr. Denis Evans of Brigham and Women's Hospital.

Abortion Rates

1.6 million abortions are performed annually. Nearly one third of pregnancies end in abortion.

Source: The Alan Guttmacher Institute.

Abortion Patients

Marital status	
Never married	63.3%
Married	18.5%
Separated	6.4%
Divorced	11.2%
Widowed	0.6%
Race	
White	68.6%
Nonwhite	31.4%
Religion	
Protestant	41.9%
Catholic	31.5%
Jewish	1.4%
No religion	22.2%
Other	2.9%

Source: Ibid.

Radon Causing Cancer

Radon is believed to cause 20,000 lung-cancer deaths a year in the United States. Radon is a colorless, odorless gas formed when uranium in soil and rocks decays into radioactive particles.

Source: Geochemist Arthur W. Rose of the Pennsylvania State University.

Gallstones for Women

Painful gallstones and kidney stones strike nearly 1.5 million people each year. Women between the ages of 20 and 60 are twice as likely to develop gallstones as men.

Source: *Good Housekeeping,* September 1989.

Kidney Stones for Men

4 out of 5 patients with kidney stones are men, usually between the ages of 30 and 50.

Source: *Good Housekeeping,* September 1989.

Aching Backs

80 percent of all Americans will have back problems sometime in their life.

Source: John Pekkanen.

Mental Fears

Almost 9 percent of the population have an anxiety disorder, but with proper therapy 80 percent of them can expect to overcome their disorder.

Source: National Institute of Mental Health.

Donating Blood

5 percent of all Americans give blood at least once a year. Of those who do, 80 percent are repeat donors.

Source: American Red Cross Gallup Poll.

Decaf Coffee Increases Cholesterol

Drinking decaffeinated coffee can raise cholesterol levels by 7 percent.

Source: American Heart Association.

Bedtime Tantrums

As many as 25 percent of children 1 to 4 years old have bedtime tantrums.

Source: Lisa Adams of Arkansas Children's Hospital.

Infertility Causes

Endometriosis causes about one third to one half of the infertility in the United States.

Source: Dr. Robert Wild of the University of Oklahoma Health Sciences Center.

Abortion Safer Than Having a Baby

Having an abortion is 7 to 25 times safer than carrying a fetus to term and does not increase a woman's subsequent risk of infertility or miscarriage.

Source: U.S. Department of Health and Human Services.

There's a Virus Going Around

10 percent of the U.S. population is affected by influenza each year.

Source: Hoffman-La Roche Flu Track Center.

Diabetes

Each day 1,500 people will be told they have diabetes. 20 percent of all Americans over the age of 55 are diabetics.

Source: American Diabetes Association.

Arthritis—Hot and Cold Climates

35 million Americans reported suffering from arthritis in 1987, with the highest incidence rate in Florida (18.1

percent) and the lowest incidence rate in Alaska (8.8 percent).

Source: Centers for Disease Control.

Prenatal Abuse

Women of childbearing age (15 to 44) are not always thinking of their child's best interest; in fact, 15 percent are current substance abusers. 44 percent tried marijuana and 14 percent used cocaine at least once during their pregnancy.

Source: National Institute on Drug Abuse.

AIDS—Mothers and Babies

70 percent of New Jersey women who abuse drugs test positive for exposure to AIDS. Those who are mothers have a 50-percent chance of giving birth to a baby with AIDs.

Source: Dr. James Oleske, Director of Pediatric AIDS Unit, Children's Hospital, Newark, New Jersey.

Don't Try It, You Might Like It

Of people age 12 and over, 37 percent of the population have tried marijuana, cocaine, or other illicit drugs in their lifetime. 14 percent used them at least once in the last month. 75 percent of the American population have tried cigarettes and 29 percent are current smokers.

Source: National Institute on Drug Abuse.

"Do You Mind If I Smoke?" "Of Course I Mind!"

Passive smoking is dangerous to your health. People living with smokers are 30 percent more likely to suffer from heart disease or heart attacks than those who live in smoke-free households. An estimated 37,000 Americans die annually from heart disease contracted from passive smoke, making it the third-leading cause of preventable death behind active smoking which kills 400,000 each year and alcohol which kills an estimated 100,000 annually.

Source: Cardiovascular Research Institute, University of California–San Francisco.

Schizophrenics

Approximately 1 percent of the population develop schizophrenia during their lives.

Source: Judith Levine Willis, Editor, FDA's *Drug Bulletin*.

Eye Problems

Black adults are more than twice as likely as whites to be legally blind or vision-impaired. 1.6 percent of blacks over 40 are legally blind vs. 0.9 percent of whites. 3.3 percent of blacks have a visual impairment that can't be corrected by glasses vs. 2.7 percent of whites.

Source: *Archives of Opthalmology*.

Odds of Dying During the Current Year

Age	White Male	Black Male
25	561 to 1	311 to 1
35	552 to 1	200 to 1
45	242 to 1	101 to 1
75	16 to 1	15 to 1

Source: Tom and Nancy Biracree. *Almanc of the American People*. Facts of File: 1988.

Antidrug Efforts

Antidrug efforts cost the federal government approximately $32 for every person in this country, totaling $7.6 billion dollars in 1990.

Source: President Bush's National Drug Control Strategy.

Drugs in the Workplace

About 9 percent of the U.S. work force show up for work each day with illegal substances in their systems.

Source: U.S. Bureau of Labor Statistics.

Becoming Addicted

About 10 percent of those who drink become alcoholics, and about 20 to 30 percent of people who try cocaine become addicted.

Source: National Institute on Drug Abuse.

Calling All Doctors

The average American consults with a doctor 5.4 times a year. Most contacts are at the doctor's office, an average of 3.1 visits per person each year. Children under age 5 have 6.7 doctor contacts a year; people between ages 65 to 74 average 8.2 contacts a year; and people over age 75 see or talk to doctors an average of 9.9 times a year.

Source: National Center for Health Statistics.

Shots for Kids

About 25 percent of preschoolers and about 33 percent of poor children go unprotected from whooping cough, measles, mumps, rubella, and polio, but 95 percent of school-age children finally get their shots.

Source: American Academy of Pediatrics.

Smoking Is Bad for a Woman's Heart

Woman smokers have a 3.6 times greater risk of having a heart attack compared with women who have never smoked. However, the risk is the same if the smoker quits for three or more years.

Source: *New England Journal of Medicine*.

Heavy Smokers Can't Smell

2-or-more-packs-a-day smokers lose 15 to 20 percent of their sense of smell. It takes 10 years for the presmoking sense of smell to return.

Source: University of Pennsylvania School of Medicine.

Obese Women Have Heart Disease

40 percent of all heart disease in women is due solely to being overweight.

Source: Boston's Brigham and Women's Hospital.

Eat Right

An estimated 35 percent of all cancers are diet-related.

Source: American Cancer Society.

Socialized Medicine Is Better, but . . .

73 percent of Americans favor having the government provide free health care but only 43 percent favor it if it means traveling farther to the doctor; only 36 percent favored it if it means waiting longer for care; and only 29 percent favor it if it means restricting their choice of health-care providers.

Source: American Society of Cataract and Refractive Surgery.

Helping the Heart

Eating less fat would reduce the average person's blood cholesterol level by at least 10 percent and cut heart disease by 20 percent nationwide. About 55 percent of U.S. adults have a cholesterol level of borderline or above.

Source: The National Cholesterol Education Program.

Bad Milk

A recent study found that 51 percent of the milk tested was contaminated with drugs used by the dairy farmers to treat their cows.

Source: Food and Drug Administration.

Too Much TV for Kids

Children ages 2 to 5 watch about 25 hours of television weekly; ages 6 to 11 watch 22 hours; and ages 12 to 17 watch about 23 hours per week.

Source: A. C. Nielsen Company.

Blood Types

The following are the percentage of Americans with their blood types:

Blood Type	Percentage
O positive	36
O negative	6
A positive	38
A negative	6
B positive	8
B negative	2
AB positive	3.5
AB negative	.5

Source: American Red Cross.

Contracting AIDS from a Blood Transfusion

In 1987, the odds of contracting HIV infection were 1 in 153,000 per unit transfused. A patient who received the average transfusion (5.4 units) had odds of 1 in 28,000.

Source: Ibid.

Pass the Blood, Bub

Approximately 2 percent of Americans were transfused in the past year. In every minute of the day, 23 units of blood or red blood cells are transfused.

Source: Ibid.

S-Stuttering

An estimated 1 to 2 percent of the population stutters, with the ratio of males to females about 4 to 1.

Source: National Council on Stuttering.

It Won't Be Long Now . . .

In 1987, 58.6 percent of all newborns were circumcised. Uncut babies (57.5 percent) now outnumber those circumcised in the west, while nearly half of all babies in the south (43.3 percent) are uncut.

Source: National Center for Health Statistics.

Good-bye Tooth Decay

Half of all U.S. schoolchildren had no cavities in 1987, compared with about 28 percent in the early seventies.

Source: National Institute on Dental Research.

Making $ Off the Tooth Fairy

4 percent of working adults in this country are missing all their teeth; however, 42 percent of senior citizens were missing all their teeth, and only 2 percent still had all 28 permanent teeth.

Source: Ibid.

Fat Women

18 to 25 percent of American women are actually overweight. The odds for a black woman are significantly

higher; 47 percent of black women are said to be overweight. About 70 percent of obesity is said to be inherited.

Source: National Association to Aid Fat Americans.

Fat Ladies Who Diet Don't Permanently Lose Weight

While about 79 percent of American women diet at some point in a given year, it is estimated that 95 percent of diets fail, meaning that these women will gain their weight back within 3 to 5 years. Of that 95 percent, some 40 percent (almost half) regain more weight than they lost.

Source: National Association to Aid Fat Women.

Medical Hypocrites

17 percent of American doctors smoke.

Source: Lewis Lapham, Michael Pollan, and Eric Etheridge. *The Harper's Index*. Henry Holt, 1987.

Aborting Unwanted Pregnancies

26 percent of American pregnancies end in abortion.

Source: *Health United States 1986*.

Dangers of the Pill

78 percent of American women think there are substantial risks involved in using birth-control pills.

Source: *Public Opinion*, February/March 1986.

Attempted Suicides

Each year, there are an estimated 600,000 suicide attempts made in the United States. Approximately 5 million living Americans have attempted to take their own life. There are 3 female attempts for every male attempt.

Source: American Association of Suicidology.

Death by Suicide

An average of 1 person every 17.1 minutes kills themselves. Each year, over 30,000 Americans commit suicide.

Source: Ibid.

Too Much Meat

Americans consumed an average of 218.4 lbs. of meat per person in 1989; in excess of a half pound of meat each day.

Source: American Dietetic Association.

Abusing the Pill

Only 11 percent of U.S. women use the pill correctly:
83% took the pills at different times
58% didn't take the pill every day
40% didn't use backup methods like diaphragms or condoms when needed
2% didn't take pills in the right order
2% borrowed pills from other women.

Source: Deborah Oakley of the University of Michigan, Ann Arbor.

Women in Nursing Homes

Female nursing-home patients outnumber males nearly 3 to 1; just under one third of all residents had no immediate family. 9 out of 10 are white. Two-thirds are 80 or over.

Source: Health Care and Policy Research.

No Sweat Workout Preferred by Women

Women represent the majority of participants in these low-impact activities:

Fitness walking	65%
Stationary biking	59%
Nordic ski machines	57%
Treadmills	55%

Source: American Sports Data, Inc., January 1990.

Odds for Getting Cancer

About 76 million Americans now living will eventually get cancer—about 30 percent. Cancer will strike in approximately 3 out of 4 families. In 1990, about 1,040,000 people will be diagnosed as having cancer, and about 510,000 will die.

Source: American Cancer Society.

Cancer Survival Rates

An estimated 50 percent of all people who contract cancer will be alive 5 years after they are diagnosed.

Source: Ibid.

Smoking Causes Cancer

It is estimated that 83 percent of lung-cancer cases are a result of smoking. This year, smoking will result in 390,000 deaths.

Source: Ibid.

Totally Blind Readers

Only about one third of the totally blind population does not use braille. About 105,000 people are legally blind in this country.

Source: American Foundation for the Blind.

Legally Blind Readers

Of the 500,000 legally blind American population, only about 15 to 20 percent use braille.

Source: Ibid.

Kids Having High Cholesterol

More than 1 in 4 U.S. children have cholesterol levels of 180 milligrams or higher.

Source: The American Health Foundation.

Need a Doctor? Go to Maryland

Maryland ranks first in physicians per capita—325 per 100,000 population.

Source: *Statistical Abstract of the United States 1987*.

Uninsured Youth

12 million children are not covered by health insurance of any kind.

Source: The United Way.

Prenatal Care a Must

Children born to women who fail to receive prenatal care are 3 times more likely to die in infancy than those whose mothers receive comprehensive care.

Source: National Association of Children's Hospitals and Related Institutions.

Children with AIDS

60 to 70 percent of children with AIDS die within 2 years of diagnosis. For every child that is diagnosed with the HIV virus, health experts estimate that another 2 to 10 children are actually infected with the virus.

Source: Centers for Disease Control.

Crack Babies

Annually, an estimated 375,000 infants are exposed to health-threatening drugs because 1 in 10 mothers takes illegal drugs during pregnancy. It costs $100,000 to take care of a crack-addicted infant during the first 3 months of its life.

Source: The United Way.

Teenage Suicides

25 percent of today's adolescent boys and 42 percent of adolescent girls say they have seriously considered committing suicide at some point in their lives. 18 percent of girls and 11 percent of boys reported actually trying to commit suicide.

Source: Ibid.

C-Sections

Women over 35 have a caesarean section rate of 40 percent compared with 22 percent among women ages 20 to 25.

Source: Mark Landon of the Ohio State University, Columbus.

Fat Self-Images

78 percent of American adult women see themselves as overweight, while only 27 percent of them actually are.

24 percent say that feeling fat has sometimes made them avoid sex.

88 percent avoid wearing a bathing suit, and 52 percent avoid slacks and shorts.

45 percent say their hips and bottom area are what they dislike the most about their bodies.

75 percent say they diet to feel better about themselves, 21 percent for health reasons, and 4 percent to please their husbands.

Source: *Family Circle*, February 1, 1990.

Who's Dieting?

1 out of 4 Americans is currently dieting.

Source: Calorie Control Council.

Tennis Elbow Anyone?

Almost one third of all tennis players, from novices to pros, end up with this classic overuse syndrome at one time or another.

Source: *Good Housekeeping*, September 1990.

Multiple Births

Twins occur in 1 out of approximately 87 births.

Triplets occur in 1 out of approximately 7,569 births.

Quadruplets occur in 1 out of approximately 658,507 births.

Quintuplets occur in 1 out of approximately 57,289,761 births.

Sextuplets occur in 1 out of approximately 4,984,109,107 births.

Source: Ashley Montagu. *Human Heredity*. Cleveland: World Publishing, 1959.

Most Chronic Disease in America

About 1 out of 8 Americans suffers sinus infections, surpassing arthritis as the most common chronic disease in the United States.

Source: National Center for Health Statistics.

4 Percent of Americans Need Help for Health Reasons

4 percent of Americans over age 15 need help performing 1 or more of the 5 daily activities as a result of a health condition lasting in excess of 3 months: doing

homework, fixing meals, managing money, getting around outside, and personal care.

Source: U.S. Census Bureau.

Girl Babies Will Outlive Boy Babies

Girls born in 1989 can expect to live 78.6 years as compared with 71.8 for boys. This averages 75.2 years for both sexes.

Source: Metropolitan Life.

The State of Your States Health

Some states are healthier to live in than others. Minnesota and Utah rank the most healthy of the 50 states, while Alaska and West Virginia are the least healthy. This study uses data from government agencies and health organizations and was based on 17 statistical measures of health. Included were such things as smoking, traffic death rates, violent-crime rates, motor-vehicle death rates per 100,000 miles driven, incidences of major infectious diseases, life expectancy at birth and access to health care.

How the States Rank

State	Score	State	Score
Minnesota	120	North Dakota	110
Utah	120	Maine	100
New Hamsphire	119	Virginia	109
Hawaii	118	New Jersey	108
Nebraska	116	Rhode Island	107
Connecticut	116	Montana	106
Massachusetts	114	Ohio	105
Wisconsin	114	Pennsylvania	104
Iowa	114	Indiana	104
Kansas	113	California	103
Colorado	113	Michigan	102
Vermont	110	South Dakota	101

State	Score	State	Score
Maryland	101	Kentucky	94
Oklahoma	100	Alabama	93
Wyoming	100	Arkansas	93
Delawarey	100	Arizona	93
Missouri	100	South Carolina	93
Washington	99	Oregon	91
Texas	99	Florida	90
North Carolina	98	New Mexico	88
Idaho	97	Louisiana	88
Georgia	96	Nevada	87
Tennessee	96	Mississippi	87
New York	96	West Virginia	85
Illinois	96	Alaska	84

Source: Northwestern National Life Insurance Company.

Cancer Risks from Various Sources Compared with Drinking Tap Water

Relative Risk	Source of Daily Exposure	Carcinogen
1.0	Tap Water, 1 liter	Chloroform
30.0	Peanut butter, 1 sandwich	Aflatoxin
4,700.0	Wine, 1 glass	Ethyl alcohol
0.3	Coffee, 1 cup	Hydrogen peroxide
2,700.0	Cola, 1	Formaldehyde
100.0	Basil, 1 g. of dried leaf	Estragole
0.4	Bread and grain products average U.S. diet	Ethylen dibromide
2,100.0	Breathing air in a mobile home, 14 hours	Formaldehyde
5,800.0	Breathing air at work, U.S. average	Formaldehyde
16,000.00	Sleeping pill, 60 mg.	Phenobarbitol

Source: "Ranking Possible Carcinogenic Hazards." *Science*, April 17, 1987.

The Older You Are, the Longer You'll Live

Americans who celebrate their 75th birthdays have an average life expectancy of 85.9 years.

Source: Metropolitan Life.

It's All in the Genes

Between 12 and 15 million Americans have a genetic disorder of one kind or another. A total of 20 million Americans are carriers of true genetic defects. 1 out of every 250 newborn babies has a genetic disorder. 1 out of every 3 babies or young children admitted to a hospital is there because of a genetic problem.

Source: Myra Vanderpool Gormley. *Family Diseases: Are You at Risk?* 1989.

Genetic Factors

It is believed that genetic factors may be involved in 25 percent of diseases.

Source: Ibid.

Like Father/Mother, Like Child

When a parent has a dominant gene for a disease, there is a 50-percent risk that each child will manifest the defect.

Source: Ibid.

Women Who Smoke Get Ulcers

1 of every 10 women who smoke can expect to develop a stomach ulcer in the next 12 years. Those women who kick the habit can reduce their risk of ulcers to nearly that of women who have never smoked. The risk is the greatest among women who smoke a pack or more of cigarettes a day.

Source: *Archives of Internal Medicine.*

Allowances May Be Dangerous to a Child's Health

Children between the ages of 4 and 12 receive more than $4.7 billion in allowances per year. Of this amount, they spend $1.4 billion (29 percent) on snacks and candy. Toys and games account for $1.1 billion; video-arcade games are $0.8 billion; movies and sports are $0.7 billion; savings are $0.5 billion; and gifts are $0.2 billion.

Source: Miami University, Oxford, Ohio.

Break the Habit . . . And Keep the Doctor Away

A 1990 study of Bank of America retirees revealed that one-pack-a-day retirees who quit smoking today could reduce next year's health bills by $2,300 each.

Source: J. Paul Leigh, San Jose State University study.

Birth Spacing

When there are fewer than 2 years between births, the risk of the second child dying in infancy is increased between 60 and 70 percent. The odds are double that the earlier child will die before age 5.

Source: Population Crisis Committee.

Low Survival Rate of American Babies

1 percent of babies born in the U.S. die before reaching their first birthday. While this is a much better rate than the 10-percent death rate in Africa, the U.S. infant mortality rate is twice as high as Japan's and is higher than many other industrial nations.

Source: U.S. Census Bureau.

Causes of Infertility

According to latest available information, 1 in 7 couples will have problems conceiving. A breakdown of causes shows:

Men - 40 Percent		Women - 40 Percent	
Sperm defects	20%	Ovulatory defects	12%
Varicocele	10%	Tubal defects	12%
Hormonal problems, duct obstructions, immunological problems	10%	Endometriosis	8%
		Immunological problems, cervical mucus, uterine defects	8%
Unexplained factors (both sexes) 20%			

Source: American Fertility Society.

Women and Heart Disease

Heart disease is the number-1 killer of women over 60.

Source: *American Health*, October 1990.

Young Couch Potatoes Risk Heart Disease

According to a study, kids who averaged 2 or more hours of television viewing a day had high cholesterol levels: 8 percent had levels about 200 milligrams per deciliter and 13 percent had levels between 176 and 199. For children age 2 and older, a level of 175 is considered high enough to warrant dietary therapy.

Source: College of Medicine, University of California–Irvine.

Teens Who Smoke

12 percent of American teens between ages 12 and 17 smoke.

Source: U.S. Department of Health and Human Services.

The Cost of Cancer in U.S.

The overall costs for cancer in the United States during the year 1990 was estimated to have been $104 billion, or $416 per person.

Source: National Cancer Institute.

Early Detection Saves Lives

Regular screening and self-examination can detect cancers of the breast, tongue, mouth, colon, rectum, cervix, prostate, and testis, as well as melanomas at an early stage when treatment is more likely to succeed. These forms of cancer represent about 50 percent of all new cancers, and of these cases, about 66 percent currently survive five years. With early detection, about 87 percent would survive.

Source: American Cancer Society.

Cancer Deaths in U.S.

Of every 5 deaths from all causes in the United States, one is from cancer. An estimated 514,000 will die of the disease in 1991, or about 1,400 people a day.

Source: Ibid.

Infertile Women

About 1 in 12 married women in the United States between ages 15 and 44 are infertile. Infertility is defined as the inability to conceive after 12 months or more of trying.

Source: 1988 National Survey of Family Growth. National Center for Health Statistics.

See You in the Nursing Home

Of the 2.2 million Americans who turned age 65 in 1990, more than 900,000 of them, or 43 percent are expected to enter a nursing home at least once. 66 percent of people using nursing homes will be women. Nearly 33 percent of these people will spend at least 3 months in a nursing home, 25 percent at least a year, and 9 percent at least 5 years.

Source: U.S. Agency for Health Care Policy Research.

Smoking and Drinking Will Stunt Infant's Growth

Babies born to women who both smoked and drank weighted 1.1 pounds less than babies whose mothers did neither. In the lower weight ranges, the extra weight could be the difference between life and death.

Source: "Does Maternal Tobacco Smoking Modify the Effect of Alcohol on Fetal Growth?" *American Journal of Public Health*, January, 1991, pp. 69–73.

Injuries: The Number One Killer of Children

Each year, American children make 16 million visits to emergency rooms for treatment of injuries. About 600,000 are hospitalized and 20,000 die. You can significantly reduce the odds of your children being injured by using seat belts and bike helmets, fencing your pools, and keeping guns out of your house.

Source: "Childhood Injuries in the United States," *American Journal of Diseases in Children*, June 1990, pp. 627–720.

Quitting Smoking Will Add Years and Pounds to Your Life

An average of 5 years after they stopped smoking, men put on 6 more pounds than those who continued to smoke and women put on 8 more pounds. 10 percent of the men and 13 percent of the women had a *major* weight gain— about 30 pounds or more. For both sexes, blacks were considerably more likely to gain weight after quitting. Smokers who were underweight were 4 times more likely than other quitters to put on 30 pounds or more. The odds of having a weight increase were greater for heavy smokers, those under 55, and women who have had children. In spite of this, the health benefits of giving up cigarettes far outweigh the extra pounds gained—even for those who put on a lot of pounds.

Source: Centers for Disease Control.

10

GAMBLING

A Bet on Roulette

In American roulette, the house has an unyielding advantage of at least 5.26 percent on every bet. (There are 36 numbers plus 2 zeros vs. 1 zero on the European wheel.)

Source: Edwin Silberstang. *The Winner's Guide to Casino Gambling*. HR & W Publishing, 1980.

Roulette Odds

Type of Bet	Payoff Odds	True Odds	House Percentage
Single number	35 to 1	37 to 1	5.26%
2-numbers	17 to 1	18 to 1	5.26%
3-numbers	11 to 1	35 to 3	5.26%
4-numbers	8 to 1	17 to 2	5.26%

NOTE: This chapter does not include sources on gambling odds that are purely mathematical.

Type of Bet	Payoff Odds	True Odds	House Percentage
5-numbers	6 to 1	33 to 5	7.89%
6-numbers	5 to 1	16 to 3	5.26%
12-numbers	2 to 1	13 to 6	5.26%
18-numbers	even	10 to 9	5.26%

Source: John D. McGervey. *Probabilities in Everyday Life*. Ballantine Books, 1986.

Baccarat

This game in which odds and play remain rigid gives the house a 1.36-percent advantage.

Source: Ibid.

Shooting Craps

If a craps player uses the best play at a craps table, the house has no more than an 0.8-to-0.6-percent advantage.

Source: Ibid.

Lucky 7

The odds of rolling a number 7 on a one-roll basis are 5 to 1. Since casinos pay 4 to 1 on this bet, the house edge is a staggering 16.67 percent. This is a real sucker's bet.

Snake Eyes

The odds of rolling a number 2 on a one-roll basis is 35 to 1. The same odds apply to the number 12. Since the house pays 30 to 1 on a 2 or 12, it has an advantage of 13.89 percent. When the house pays 30 for 1 (the same as 29 to 1), it has a 16.7-percent edge!

36 Combinations

There are 36 different combinations with a pair of dice. The following are the true odds on a onetime roll against each of the numbers listed below:

3 and 11	17 to 1
4 and 10	11 to 1
7	5 to 1
5 and 9	8 to 1
6 and 8	31 to 5

Odds of Numbers Against a 7

The odds against repeating the following number before a 7 are:

4	2 to 1
5	3 to 2
6	6 to 5
8	6 to 5
9	3 to 2
10	2 to 1

The Pass Line

90 percent of all players in a gambling casino will play the pass line.

Source: Edwin Silberstang. *The Winner's Guide to Casino Gambling.*

7 Come 11

The comeout roll of a 7 or 11 is an automatic winner. There are 8 ways to roll a 7 or 11, which means the odds are 9 to 2 against either appearing on a single roll.

Immediate Losers

On a comeout roll, a 2, 3, or 12 is an automatic loser. There are 4 combinations out of 36 by which these numbers can appear, so the odds are 1 in 8 times that the player will be an immediate loser.

Free Odds

Once a bettor has made a bet on the dice, a gambling casino will allow him or her to bet against the 7 on a 4 or 10 and receive a payoff at 2 to 1—the exact odds; with the 5 and 9, the payoff is 3 to 2, and with a 6 or 8, the payoff is 6 to 5. These bets against the 7 are referred to as free odds because the house has no advantage over the player. When a gambling house permits a player to double his wager with these free odds, the odds in favor of the house are only 0.6 percent, the lowest on any casino game (with the exception of blackjack when a player "counts the cards").

Place Bets

After the comeout roll at the craps table, a player can bet on the place numbers 4, 5, 6, 8, 9, and 10. The chart below shows the house payoffs, the correct payoffs, and the house advantage (the difference between what the player receives and the true odds):

Number	Payoff	Correct Payoff	House Edge
4	9 to 5	2 to 1	6.67%
5	7 to 5	3 to 2	4.0%
6	7 to 6	6 to 5	1.52%
8	7 to 6	6 to 6	1.52%
9	7 to 5	3 to 2	4.0%
10	9 to 5	2 to 1	6.67%

The casino has too big an edge on the 4, 5, 9, and 10, so over a period of time, such odds are too high to overcome. Even the 6 and 8 bets are poor odds in comparison with the free-odds bets.

Field Bets

A player at the craps table can make a field bet, which means he will win when the dice show a 2, 3, 4, 9, 10,

11, or 12. If a 2 or 12 is thrown before the 5, 6, 7, or 8, he receives a 2-to-1 payoff; the 3, 4, 9, 10, and 11 pay even money. In effect, he has 18 chances (vs. 16 as a result of the numbers 2 and 12 having a 2-to-1 payoff) of winning against 20 ways that the 5, 6, 7, or 8 can appear on the dice. This gives the house a 5.5-percent advantage. In casinos where the 2 and 12 are paid at 3 to 1, the house edge is reduced to 1.7 percent.

The Big 6 and the Big 8

The odds favor the gambling casino by 9.9 percent when the player places a bet on 6 and 8 (a.k.a. the big 6 and the big 8) at the craps table.

The Horn Bet

This bet at the craps table is placed on the numbers 2, 3, 11, or 12 against the rest of the numbers on a onetime roll. The odds are 5 to 1 against these 4 numbers appearing on a single toss of the dice.

Hardways

In craps, the 4-point numbers, 4, 6, 8, and 10 must appear as a pair (i.e., 2 2s = 4) on a single roll of the dice against the 7 or the easy way. The hard 4 and hard 10 have 8-to-1 odds and the payoff is 7 to 1, giving the house a 11.1-percent edge. The hard 6 and hard 8 have 10-to-1 odds, and the payoff is 9 to 1, giving the house an edge of 9.09 percent. Again, these are sucker bets.

Counting in Blackjack

Counting cards gives an advantage to the player over the dealer because the player has options but the dealer does not. The dealer must hit when his count cards are less than 17. Basically this means that when there are many 10s left in the deck, the player who counted has an advantage because he can stand with, say, a 15 and the dealer has a 6 or less showing. This is true because a 10 card will bust the dealer. Additionally, with many 10 cards and aces remaining, the player is paid 3 to 2 with a blackjack while the dealer gets even money. Also, when the dealer has an ace and you know how many 10 remain, you can calculate whether it is to your advantage to buy "insurance."

Buying Insurance at the Blackjack Table

Assuming on the first hand dealt it's you against the dealer, there are no 10s out, and the dealer has an ace showing. Your cards are a 6 and 9. There are 16 10s in a deck of 52 cards, which means there are 36 other cards. When you subtract the dealer's ace (his card that's showing) and your 2 cards, there are 33 non-10 cards against 16 10s. Since you can buy insurance against the dealer having a blackjack and win 2 to 1, the dealer's odds are 33 to 16 against your 32 to 16—an edge that means it's a sucker's bet to buy the insurance. Now let's say your hand is a king and queen and the dealer shows an ace. Now the odds are in your favor because there are 35 non-10s left in the deck versus 14 10s and the insurance bet pays off at 2 to 1.

Keno

The house has an edge in keno that's never less than 22 percent and can be as high as 40 percent.

Source: Edwin Silberstang. *The Winner's Guide to Casino Gambling*.

Chuck-A-Luck

This game is played with a device that resembles an hourglass with a bird cage on each end. Three oversized dice with numbers 1 through 6 on each cube are put in the bottom cage, and you place your bet on a number (1 through 6). When the cage is flipped, even money is paid if your number appears once, 2 to 1 if your number appears twice, and 4 to 1 if it appears on all 3 dice. The house has a 7.87-percent edge over the player.

Source: Ibid.

One-armed Bandits

The odds are never in your favor that you'll beat a slot machine because the house is in business to make money so it programs the machine to win. Naturally! This means that what the odds are depends on where you gamble. In Las Vegas, the worst places to play are the grocery stores, drugstores, restaurants, supermarkets, and other establishments that are in a business other than gambling. This includes McCarran Airport, where some machines pay out no more than 50 percent of the coins taken in. Las Vegas's downtown casinos only have a 10-to-12-percent edge, and the Strip casinos operate with an edge of an estimated 18 percent.

Source: Ibid.

The Big 6

Known as the jumbo dice wheel, this giant wheel of chance is 5 feet in diameter. It contains 54 spaces around the rim, each of which shows 1 side of 3 dice bearing different combinations of the numbers 1 through 6. The house enjoys an advantage of 22.22 percent.

Source: John Scarne. *Scarne's Guide to Casino Gambling*. Simon and Schuster: 1978.

Casino Tables

51 percent of all gambling dollars in a casino are played at tables.

Source: "Gaming and Wagering." *Business*, July 1987.

Roulette-Wheel Winnings

7 percent of American casino-game winnings are made at the roulette wheel.

Source: Ibid.

Blackjack Winnings

49 percent of American casino-gambling winnings occur at the blackjack tables.

Source: Ibid.

Winning the Lottery

The odds of picking 6 numbers in a lottery with a field of 44 numbers is 7,059,052 to 1. When the field is expanded to 53 numbers and the player can buy 2 sets of 6 numbers, the odds of winning are 11,478,740 to 1.

Source: State of Ohio Lottery.

5 for 5 Out of 100

The odds of picking 5 numbers in a lottery with a field of 100 numbers is less than 1 in 40 million; the chance of picking 4 numbers would be about 1 in 100,000.

Source: John D. McGervey. *Probabilities in Everyday Life.*

Practically Everybody Plays the Lottery

79 percent of Americans have bought a lottery ticket.

Source: Mel Poretz and Barry Sinrod. *The First Really Important Survey of American Habits.* Price Stern Sloan: 1989.

Betting with the Bookie

When a bookmaker realizes a profit by receiving 11-to-10 odds from the bettor, no matter which side is bet, he has an edge of 4.54 percent; each time he takes in $21

($11 on each side), he gives back $21 to the winner and keeps $1.

Parlays

When you parlay 2 bets simultaneously, a bookie will give you odds of a 2.4 to 1 on 2 even propositions (on each of which you would normally give the book 11-to-10 odds). The odds are 3 to 1 against you winning, thereby giving the bookie a 15-percent edge.

Source: John D. McGervey. *Probabilities in Everyday Life.*

Football Cards

Let's assume that the correct spot is placed on each game, and 10 games are listed on a football card from which you can select 3 teams. If your selection is right, the payoff is 5 to 1 and you lose if there are any ties. Even if ties were impossible, you would be fighting an edge of 25 percent. If you try to pick 10 winners, your payoff is 100 to 1—but the true odds are 1,023 to 1.

Source: Ibid.

Odds of Improving Poker Hand

Holding Before the Draw	Number of Cards to Be Drawn	Odds of Improving Hand
1 pair	3	2, 5 to 1
2 pair	1	11 to 1
3 of a kind	2	9 to 1
4 flush	1	4, 5 to 1
2-sided straight	1	5 to 1
Straight	—	—
Inside straight	1	11 to 1

Source: Edwin Silberstang. *The New American Guide to Gambling and Games.* New American Library, 1987.

More Odds Against Improving a Hand in Draw Poker

Cards Held Before Draw	Number of Cards Drawn	Improving to	Odds Against Drawing
1 pair	3	2 pair	5 to 1
		3 of a kind	8 to 1
		Full house	97 to 1
1 pair with the ace kicker	2	Aces up	7.5 to 1
		Any other pair	17 to 1
		3 of a kind	12 to 1
		Full house	119 to 1
2 pair	1	Full house	11 to 1
Three of a kind	2	Full house	15.5 to 1
		4 of a kind	22.5 to 1
Open-ended 4 straight	1	Straight	5 to 1
1-sided straight or inside staight	1	Straight	11 to 1
4 flush	1	Flush	4.2 to 1
4-card Straight Flush, Open Ended		Straight Flush	22.5 to 1
		Any improvement	2 to 1

Source: Ibid.

On- vs. Off-Track Betting

89 percent of horse-racing bets are on-track and 12 percent are off-track. The average amount per winning ticket is $8.80.

Source: Thoroughbred Racing Association.

Casinos Bring in Big Bucks

The average daily revenue for Atlantic City's top 5 in 1989:

Trump Plaza	$956,438
Caesar's	$913,014
Harrah's Marina	$894,808
TropWorld	$891,671
Bally's Park Place	$889,967

Source: Casino Association of New Jersey.

Bridge-Game Combinations

The number of different hands that can be dealt to a bridge player is 635,013,559,600.

Source: American Contract Bridge League. *The Official Encyclopedia of Bridge.* Crown, 1987.

Being Dealt a Yarborough

The odds of being dealt a Yarborough, a dreaded hand in bridge consisting of no card higher than a 9 are 1,827 to 1.

Source: Ibid.

The Perfect Bridge Hand

The odds against receiving the perfect bridge hand, one that will produce 13 tricks in no-trump regardless of the composition of the other hands, are 169,066,442 to 1.

Source: Ibid.

Receiving a Standard Point Count to Open

The probability of receiving fewer than 12 high-card points in a dealt bridge hand is 65.183 percent, so disregarding distributional values, about two-thirds of all hands would generally be considered too weak to open at one level.

Source: Ibid.

Other Bridge Odds

The probability of a hand in the 15-to-17-point one no-trump range is 10.097 percent. The odds that you will receive exactly your share of the high-card points (10) is 9.405 percent.

Source: Ibid.

11

MONEY MATTERS

Being Audited

1 percent of American income-tax returns are examined by the IRS.

Source: *Statistical Abstract of the United States 1989*. Bureau of the Census. U.S. Department of Commerce.

Paying Cash

12 percent of American households pay their bills exclusively in cash.

Source: Lewis Lapham, Michael Pollan, and Eric Etheridge. *The Harper's Index*. Henry Holt; 1987.

Small Investors

45 percent of American stockholders have a portfolio worth under $5,000.

Source: *Shareownership 1985*. New York Stock Exchange, Inc.

Taxing Togetherness

48 percent of American income-tax returns are filed jointly.

Source: *Statistical Abstract of the United States 1989*.

Silent Partners

64 percent of American housing units are mortgaged.

Source: Ibid.

It Pays to Serve Yourself

In 1989, those customers who pumped their own gasoline saved an average of 2 cents a gallon. This means an average of 1 weekly fill-up of 10 gallons saves $114.20 a year.

Source: American Automobile Association.

IRS Odds

If you make less than $25,000 and file an uncomplicated return, the odds are less than 1 in 100 that you will be audited.

If your income is $50,000 or more, those odds increase to about 2 in 100.

Source: Internal Revenue Service.

1 in 1,000 of the Rich Don't Pay Taxes

In 1987, of the 529,460 couples and individuals who reported income in excess of $200,000, 595 paid no tax, which is slightly more than 1 in 1,000. The earnings of these 595 averaged $490,000.

Source: Ibid.

How Much Do Americans Save?

12 percent save nothing
22 percent save 1 to 9 percent
32 percent save 10 to 19 percent
18 percent save 20 percent or more
16 percent don't know
The average American saves 4 percent of his annual

income. These figures increase to 14 percent between the ages of 45 through 64.

Source: *Money*, April 1989.

Gasoline by Credit Cards

31 percent of Texaco customers purchase gasoline and pay by its credit card.

Source: Texaco Oil Company.

Nothing but the Best for Mother

The average amount spent per gift for Mom on Mother's Day is $25.95.

Source: The Gallup Organization in a 1989 survey conducted for *Gift and Stationery Business*.

The IRS Gives, Too

70 percent of all tax filers receive refunds.

Source: Internal Revenue Service.

Average Income for Family of 4

$30,000 is the average earnings for a family of 4.

Source: Tax Foundation.

Who Tips How Much

People who eat in Italian restaurants tip the highest—an average of 15.1 percent. At seafood restaurants, the tips average 14.1 percent. Mexican restaurants average 14.8 percent; Chinese, 15 percent; general menu (58 percent of all meals), 15.2 percent.

Source: University of Illinois survey (hired by IRS).

Corporate Payouts

67 percent of American corporate profits are distributed as dividends.

Source: *Country Report: USA*. Fourth Quarter 1987. The Economist Intelligence Unit.

Paying by Check

79 percent of all American families have checking accounts.

Source: Statistical Abstract of the United States 1989.

Money for the Arts

12 percent of the money businesses give to nonprofit organizations goes to the arts.

Percentage of money given to the arts:

Museums	16%
Symphony orchestras	16%
Theater	12%
Dance	8%
Opera	8%
Public radio/TV	8%
All others	32%

Source: Business Committee for the Arts, Inc., National Study.

High Film Costs

In 1989, the average film cost was a record $23.5 million. Moviegoers bought $1.1 billion worth of tickets.

Source: The Motion Picture Association of America.

Standard Taxpayer Deductions

In 1988, only 29.2 percent of returns filed were itemized.

Source: Internal Revenue Service.

Health-Care Costs

The cost of health care per employee in 1989 was $2,748.

Source: A. Foster Higgins and Company.

Time for Savings

In 1989 we saved 5.5 percent of our after-tax income. Best year for savings: 1944—we saved 25.1 percent. Worst year for savings: 1933—we saved 3.6 percent.

Source: U.S. Department of Commerce.

Unfair Federal Taxes

Low benefits for poor; high for rich. For the richest 20 percent, rates dropped and incomes grew 31 percent during the 1980s. The poorest 20 percent made 3.2 percent less money, and their tax rate climbed.

Source: *USA Today*, February 21, 1990.

Grocery Expenditures

Average weekly grocery cost per person:
Size of household:

1 person	$40
2 people	$31
3 to 4 people	$25
5 or more	$19

Larger families spend less per person on groceries.

Source: Food Marketing Institute.

Where Do Our Tax Dollars Go?

How the federal government will spend our tax dollars in 1990:

Income Security	33¢
Defense	24¢

Interest on federal debt	14¢
Health	13¢
Education	3¢
Transportation	2¢
Veterans' benefits	2¢
Environment	1¢
All other	8¢

Source: Tax Foundation.

Faulty CPAs

Money magazine called the IRS 100 times and asked 10 questions 10 times each. It asked 50 tax preparers to do a sample tax return for a family of 4. The results were printed in its March 1990 issue.

Only 2 in 50 tax preparers filled out an error-free return.

The answer to how much tax was owed ranged from $9,806 to $21,216. *Money* says the right answer was $12,038.

Low Pay for Day-Care Workers

The average salary for a day-care worker with 5 years or less experience is $14,460 a year.

Source: National Association for the Education of Young Children.

High Day-Care Tuition

The average yearly tuition for 4-year-olds attending a day-care center is $3,648.

Source: Ibid.

High Cable Costs

The average monthly price of basic cable in 1990 is $16.33, up from $8.46 in 1982.

Source: *USA Today*.

Pregnancy Costs High

The average cost for a pregnancy in 1989 was $4,334, up 25 percent from 1986.

Source: Health Insurance Association of America.

Long-Distance Communication

The average cost of a 3-minute long-distance phone call is 70¢. (Daytime, 25-mile calling range, intrastate.)

Source: National Utility Service 1990 Survey.

Retirees Income

Social Security	38%
Earnings from investments	28%
Earnings from a current job	16%
Company pensions	14%
Other sources	4%

Source: "Your Money." *Good Housekeeping.* September 1989.

Expensive Attire

The average American spent $676 on clothing and shoes in 1989.

Source: *Family Economics Review,* 1990.

Household Incomes

The median household income in 1989 was $58,800.

Source: The Chicago Title and Trust Company.

Kids in Poverty

23 percent of all children under the age of 6 are living in poverty. 19 percent for children 6 to 17 years old.

Source: National Center for Children in Poverty and U.S. Census Bureau.

Impoverished Adults

11 percent of Americans aged 18 to 64 live in poverty. 12 percent of American adults aged 65 and older live in poverty.

Source: Ibid.

Paying Cash for Remodeling

67.6 percent of homeowners pay cash for remodeling. 25.3 percent get a home equity loan; 16.6 percent get the money from other sources.

Source: Professional Builder Survey.

Home Prices

The National Association of Realtors surveyed 83 metropolitan areas and found:

Lowest median-priced home—Peoria, Illinois: $47,200.
Highest median-priced home—Honolulu, Hawaii: $280,900.
National median price was $92,800 in 1989.

Source: National Association of Realtors.

Distribution of Income: A Big Gap in Incomes

In 1987, 5 percent of Americans captured 16.9 percent of aggregate income. By contrast, the poorest fifth of American families—a group 4 times the size of the richest group—earned 4.6 percent of the national income.

Source: Frank Levy of the University of Maryland.

2 Persons Live Cheaper Than One in Single Household

2-person households spend 68 percent more than 1-person households.

2-person households spend 53 percent more on housing

and 48 percent more on clothing, but spend the same amount of money on alcohol.

Source: U.S. Bureau of Labor Statistics: 1988.

Working Teens

36 percent of high-school girls work compared to 34 percent of high-school boys.

Source: *Good Housekeeping*, November 1989.

Tax Refunds

Individual tax refunds averaged $921 in 1988.

Source: Internal Revenue Service.

Income-Tax-Return Errors

Error rates are approximately 5.5 percent with electronic filing, compared with 21 percent for manual returns.

Source: Ibid.

Charitable People

Where the money comes from for the United Way:

Corporate and small-business employees	51.3%
Nonprofit and government employees	12.5%
Professionals	2.6%
Noncorporate foundations	1.4%
Other	6.1%
Corporations	23.4%
Small Businesses	2.7%

Source: The United Way.

Personal Donations

Households with an income of lower than $10,000 gave an average of 2.8 percent of their incomes to charity, whereas those who make between $50,000 and $75,000 gave 1.5 percent of their incomes. Individuals with incomes over $100,000 gave 2.1 percent of their incomes.

Source: The Gallup Organization.

Rising Prices

In 1989, the inflation rate was over 5 percent.

Source: The United Way.

Where the Money Goes

Donations given to the United Way go towards:

Health	20.2%
Family services	21.7%
Income and jobs	5.0%
Food, clothing, and housing	9.1%
Community development	5.9%
Youth and social development	17.5%
Day care	6.8%
Public safety	6.1%
Education	3.2%
Other	4.5%

Source: Ibid.

Beneficiaries of Direct Contributions

Religion	47%
Health and human services	19%
Education	9%
Arts	6%
Public	3%
Other	15%

Source: American Association of Fund Raising Counsel: 1989.

School-Nurse Salaries

The average school nurse makes $25,300 a year.

Source: National Association of School Nurses.

What an Ex-Wife Gets

Length of Marriage	Percent of Assets
5 years or less	0% to 25%
5 to 10 years	15% to 36%
10 to 15 years	33% to 45%
15 years or more	45%

Source: Raoul Felder, American Bar Association.

Financing the Future

Almost one third of the people surveyed don't spend time on personal financial planning. Hours spent per month:

0 hours	32%
11 plus	11%
Don't know	12%
4 to 10	16%
Up to 3	29%

Source: Oppenheimer Management Corporation.

Few Coupons Redeemed

The average face value of a grocery coupon was 38¢ in 1988. Of the $84 billion worth of coupons only 3.2 percent were redeemed or $2.7 billion worth.

Source: Manufacturers' Coupon Control Center.

Budget Deficit

61 percent of Americans worry that their income will not be enough to cover their expenses and bills.

Source: The Gallup Organization.

Charge!

The total number of credit cards in the United States topped the 1-billion mark in June 1990. That's 4 cards for every man, woman, and child.

Source: Nilson Report.

Pay by Plastic

$28 out of every $100 spent on consumer goods and services is charged to a credit card in the United States.

Source: Ibid.

Your Money Doesn't Necessarily Go Where You Go

26 percent of every dollar spent on airfares, hotels, and car rentals is used for federal and local taxes, travel-agent commissions, credit-card fees, computer-reservation-system charges, and freebies to large corporate clients. The biggest fee is travel-agent commissions, which absorb 13 percent of every travel dollar.

Source: Runzheimer International.

The 1.6-percent Richest Americans

28.5 percent of the U.S. personal wealth is in the hands of 3.3 million people. The IRS reports that these people had assets of $500,000 or more; their total holdings were $4.3 trillion, representing a combined net worth of $3.8 trillion in 1986, the latest year that figures were available. The Gross National Product in 1986 was $4.1 trillion.

Source: Internal Revenue Service.

The Biggest Individual Tax Deductions

The federal government estimates the following tax deductions for fiscal 1991:

Mortgage interest on owner-occupied homes	$46.6 billion
State and local income, personal property taxes	$21.9 billion
Charitable contributions	$16.2 billion
Property tax on owner-occupied homes	$12.4 billion
Medical expenses	$3.0 billion

Source: Joint Tax Committee, U.S. Congress.

Who's Deducting What?

	Percentage Taking Deduction for				
Income	Some Type	Real Estate Taxes	Medical Costs	State & Local Taxes	Charity Contributions
0 to $50K	24%	18%	7%	18%	19%
$50 to 100K	67%	65%	7%	60%	67%
$100 to $200K	90%	85%	4%	79%	89%
$200K and up	93%	83%	2%	76%	90%

Source: Ibid.

Those Who Tithe

Relatively few churchgoers tithe, or donate a tenth of income, although an estimated 10 percent of the members belonging to evangelical churches do.

Source: Southern Baptist Convention.

Who Says the Rich Don't Pay Taxes . . . ?

The top 1 percent of all taxpayers in the United States, those with incomes above $157,136, paid 27.6 percent of all income taxes in 1988. These taxpayers took in about 21 percent of all income in 1988.

Sources: U.S. Treasury and Internal Revenue Service.

American Affluence

9 percent of American households are considered to be affluent. There are 8.4 million families with incomes of $75,000 or more. Over half are moderately affluent, meaning their incomes are between $75,000 and $100,000. The other 43 percent, or 3.6 million households, are among the very affluent, with incomes of $100,000 or more. Interestingly, 21 percent of the very affluent households are likely to be traditional 1-earner couples vs. 14 percent of the moderately affluent.

Source: U.S. Census Bureau; 1989.

Affluence Among the Sexes

Of the 3.8 million individuals with incomes of $75,000 or more, 87 percent are men. Of the affluent men, 52 percent have incomes of $100,000 as compared with 46 percent of the affluent women.

Source: Ibid.

Married Men Make More Money

Married men earn an average of 30.6 more than unmarried men.

Source: University of Michigan survey.

Kids' Pocket Money

Each year, American children have $230 in pocket money, which is more than the half-billion poorest people in the world have.

Source: Worldwatch Institute.

The National Debt Is Your Debt

In 1990, the national debt, in round numbers, stood at $3.25 trillion. That breaks down to $13,000 per capita, or $52,000 for the average family. $3.25 trillion in singles would be enough bills to cover the areas of Delaware and Maryland. With 94 million households, the average debt per household is $34,574. Based on the average household income estimated at $48,840, the national debt is 71% of annual personal income.

Source: American Citizens Committee on Reducing Debt.

Average Deductions on Tax Returns

The following are average deductions that taxpapers at various income levels wrote off on their returns for 1988, the most recent year for which figures are available:

Adjusted Gross Income	Type of Deduction		
	Medical	Taxes	Charitable
$50,000–$75,000	$ 4,281	$ 3,853	$1,524
$75,000–$100,000	5,580	5,659	2,083
$100,000–$200,000	12,967	8,725	3,402
$200,000–$500,000	39,542	18,432	6,913

Source: Research Institute of America; *The Wall Street Journal.*

Those Golden Years Without Gold

95 percent of all Americans who reach age 65 cannot afford the luxury of financial independence. 22 percent must continue to work, 28 percent rely solely on welfare or social security, and 45 percent are dependent on their family. Only 5 percent are financially independent enough to meet their needs, and only 1 percent have enough financial independence to enjoy the same standard of living they had prior to retirement.

Source: *Compass readings*, January, 1991.

Old Age and Poverty—A Poor Combination . . .

More than 62 percent of Americans age 65 and older have less than $6,000 annual income, 79 percent have total assets less than $35,000, and only 8 percent receive income which exceeds $15,000 per year. These income figures include social security benefits.

Source: Ibid.

Our Poor Great-Grandparents

The average American is 4½ times richer than his or her great-grandparents.

Source: Worldwatch Institute.

12

AMERICAN THINKING

Business-Trip Experiences

4 out of 10 people think that traveling for business is an enjoyable experience; the rest think it is either a neutral or bad experience, and 16 percent hate it.

Source: *The Wall Street Journal* Centennial Survey.

Reviewing the Boss

Percentage of people who feel their boss treats them fairly and honestly:

Almost always	50%
Most of the time	29%
Occasionally or hardly ever	16%
Don't know	5%

Source: *Accountants on Call*, Gallup survey of 667 adults.

Reasons Why People Buy by Mail

Items not available in store where I shop	54%
Convenience	44%

Had a chance to think about it more
    carefully before buying   37%
Buying by mail saves money   36%

Source: Joseph Castelli and François Christen. *VALS Looks at Direct Marketing*. 1986.

Reasons Why People Don't Buy by Mail

You can't be sure of what you're getting   75%
It's more difficult to get problems
    straightened out   60%
More difficult to get written guarantee   55%
The companies are less trustworthy   49%

Source: Ibid.

Handgun Control

62 percent of Americans believe laws regarding the sale of handguns should be made more restrictive.

Source: U.S. Department of Justice.

Immoral Kids

About 25 percent of all students in grades 5 through 12 think they can experiment with drugs without abusing them. 33 percent of students say drug-education programs are working.

Source: *Scholastic Magazine*.

Dissatisfied Patients

Only 54 percent of patients think that doctors diagnose their patients properly. Only 46 percent of patients think their doctors do a good job listening to their patients. 22 percent of patients think their doctors charge them fairly.

Source: From poll of 1,012 adult Americans taken for TIME/CNN on April 4–5, 1989, by Yankelovich Clancy Shullman.

High Cost of Donating Blood

44 percent of Americans believe that donating blood can lead to infection with the AIDS virus when, in fact, giving blood does not put one at risk.

Source: American Red Cross.

Employee Health Benefits

7 out of 10 employees believe their employers should provide health-care benefits for their families, but 9 of 10 employers say they plan to reduce or eliminate coverage.

Source: Northwestern National Life Insurance Company.

Women Dislike Foreign Investment

About 57 percent of women think foreign investment in the United States is bad for the U.S. economy and U.S. workers, compared with about 45 percent of men.

Source: *Nightly Business Report*.

Beliefs on Drug Testing

97 percent of Americans say that drug testing is appropriate under certain circumstances and 85 percent say some form of drug testing helps deter drug abuse.

Source: Institute for a Drug Free Workplace.

Drug-Abuse Policies

82 percent of Americans support disciplinary actions against workers violating drug-abuse policies.

Source: Ibid.

Making Investments

Percentage of people who feel they are knowledgable on how to make each of the following investments:

U.S. Savings Bonds	74%
CDs	64%
Money fund	50%
Stock	41%
T-Bills	39%
Mutual funds	38%
Corporate bonds	33%
Tax-exempt bonds	28%

Source: *The Wall Street Journal* Centennial Survey.

Few Trust Stockbrokers

Who people trust most to give good advice on investing a $10,000 windfall:

Commercial bank	35%
Savings and loan	24%
Credit union	15%
Stockbroker	11%
Insurance company	5%
Don't know/others	10%

Source: Ibid.

Better Than Expected

Overall, Americans have had a better life than they expected. 36 percent say their standard of living is better; 37 percent think their work and career are better, and 44 percent say their family life is better than they expected when they were younger.

Source: Louis Harris and Associates, Inc.

Views on Fast Foods

76% of Americans think that the employees at fast-food restaurants are helpful and friendly. 65 percent think the sandwiches are freshly made.

Source: Wendy's/Gallup, survey of 1,029 fast-food consumers.

Earthy People

About 85 percent of Americans are interested in buying convenience foods in environmentally safe packaging and about 77 percent say they would be willing to pay a little more for these products.

Source: *Good Housekeeping* Reader Poll.

Confidence in Industries

Americans have the most confidence in the pharmaceutical industry, with 31 percent of people expressing their confidence. Next are banks with 30 percent, automobile 29 percent, food/grocery 29 percent, and appliances 27 percent. Americans have the least confidence in the airline industry, with 43 percent expressing their low-confidence level. Insurance 27 percent, oil and gas 22 percent, and stockbrokers 22 percent.

Source: *The Wall Street Journal* Centennial Survey.

Satisfied Americans

Percentage of people, in 1988, who were satisfied with the following:

Family life	94%
Health	88%
Free time	87%
Housing	87%
Standard of living	85%
Job	76%
Household income	69%

Source: The Gallup Organization.

Communication Breakdown

When asked about the quality of their relationship with their husband/wife/romantic partner:

Excellent	40%
Good	47%
Fair	10%
Poor	2%
No opinion	1%

Source: Ibid.

Remarry?

When asked if they would remarry the same partner again:

Would	88%
Would Not	8%
Not Sure	4%

Source: Ibid.

Health Over Wealth

Given only 1 choice, what is the most important thing in life—to be creative, famous, powerful, wealthy, or healthy?

Only 1 percent said power, 5 percent wealth, 6 percent creativity, 8 percent success, 10 percent a happy marriage, 16 percent to help others, and 50 percent of the survey said their health is the most important thing in life.

Source: *Los Angeles Times* Survey.

A Scary Short Walk

43 percent of Americans say that they would be afraid to walk alone at night within 1 mile of their home.

Source: The Gallup Organization.

Serious Problems

When asked what are the most important problems facing this country today, 34 percent said economic problems, 27 percent drug abuse, 10 percent poverty, 6 percent crime, 5 percent moral decline, 4 percent the environment, 4 percent international problems, and 3 percent education.

Source: Ibid.

Life After Death?

66 percent of Americans feel there is life after death.

Source: "America in the Eighties." R. H. Bruskin Associates Market Research, 1985.

Enjoying Work

68 percent of Americans are generally satisfied with their job.

Source: Louis Harris and Associates, Inc. "The Nuprin Pain Report." Bristol-Myers Products: 1985.

Does Heaven Exist

84 percent of Americans believe that heaven exists.

Source: Lewis Lapham, Michael Pollan, and Eric Etheridge. *The Harper's Index.* Henry Holt, 1987.

God?

95 percent of Americans believe in God.

Source: Louis Harris. *Inside America.* Vintage, 1987.

Plastic Surgery?

99 percent of American women would change something about their looks if they could.

Source: Ibid.

More Crime

53 percent of Americans think there is more crime in their area than there was a year ago.

Source: The Gallup Organization.

Welcome to My Neighborhood

Here's how Americans felt about whom they would and would not want as a new neighbor in 1989:

Potential Neighbors	Would be Welcome	Not Welcome	Not Sure
Catholics	94%	3%	3%
Protestants	92%	5%	3%
Jews	91%	5%	4%
Blacks	83%	12%	5%
Hispanics	78%	16%	6%
Vietnamese	75%	18%	7%
Unmarried Couples	71%	23%	6%
Religious Fundamentalists	58%	30%	12%
Religious sects, cults	31%	62%	7%

Source: Ibid.

13

ODDS AND ENDS

How Accurate Are the U.S. Census Takers?

It's estimated that the census—at 226,545,805 homes counted in 1980—was undercounted by 1.4 percent.

Source: U.S. Census Bureau.

How Long Will It Take to Decompose?

It takes 2 to 4 weeks for a traffic ticket to decompose. A cotton rag takes 1 to 5 months. A plastic six-pack ring takes more than 450 years. An aluminum can can take between 200 and 500 years.

Source: Washington Citizens for Recycling.

What's on TV for Children

Each day, the average American child will watch three and a half hours of television. During this time, he or she will view 33 acts of violence and 38 sexual references and innuendos.

Source: American Academy for Pediatrics.

Happy Birthday

Every day, 673,693 Americans have a birthday, and 3 million Americans will purchase birthday presents.

Source: Tom Heymann. *On An Average Day*. Ballantine Books, 1989.

Weather Forecasting—by the Old Farmer's Almanac

Timothy R. Clark, president of the *Old Farmer's Almanac*, claims an 80-percent accuracy in weather forecasting.

Source: *Columbus Dispatch*, October 4, 1989.

Highest and Coldest Temperatures

Juneau, Alaska, is the coolest town in the nation, with an average temperature of 40 degrees. It is also the wettest. Phoenix has the highest average temperature: 71.2 degrees.

Source: *Statistical Abstract of the United States 1987*. Bureau of the Census. U.S. Department of Commerce.

In-Shore Fishing

51 percent of American fish catch is made within 3 miles of shore.

Source: Ibid.

Paid by the Hour

59 percent of American workers are paid by the hour.

Source: Ibid.

Softbound Books

64 percent of American books are softbound.

Source: Ibid.

Lead Air Pollution

87 percent of American lead air pollution is produced by motor vehicles.

Source: Ibid.

Protestant Churches

91 percent of American churches are Protestant.

Source: Ibid.

Super Savers

91 percent of American airline passengers use discount fares.

Source: Ibid.

Telephone Service

91 percent of American households have telephone service.

Source: Ibid.

Paid Leave for Jury Duty

91 percent of American employees of medium and large firms get paid leave for jury duty.

Source: Ibid.

Tunes at Home

99 percent of American households have at least 1 radio.

Source: Ibid.

Bigger Than Texas

16 percent of America is Alaska.

Source: Ibid.

A Nation of Trees

29 percent of America is forest land.

Source: Ibid.

The Nation's Biggest Landowner

32 percent of America is owned by the government.

Source: Ibid.

Garbage

37 percent of American garbage is paper and paperboard.

Source: Ibid.

Clever Foreigners

44 percent of American patents are issued to residents of other countries.

Source: Ibid.

Technophobia

Of the 68 million households that own VCRs, about 80 percent of the owners cannot program theirs to record.

Source: Keith Morrison. *Real Life, with Jane Pauley*. NBC Television.

Deadly Storms

Deaths from storms totaled 276 between September 1987 and August 1988. Lightning killed 72, floods killed

35, tornadoes 33, thunderstorm winds 22, and high winds 17. August had the greatest number of deaths by storms.

Source: National Climatic Data Center.

Dangerous Storms

From September 1987 through August 1988, 1,909 people were injured in storms. Tornadoes injured 847, lightning 343, and thunderstorm winds 328. July had the greatest number of injuries from lightning, and November had the greatest number of injuries from tornadoes.

Source: Ibid.

Buy a Book

Approximately 93 million people visit a bookstore at least once a month and about half purchase a book.

Source: American Booksellers Association.

Environmental Concern

Over the last 15 years, concern about the environment has nearly doubled. In 1989, 55 percent of Americans think environmental laws haven't gone far enough.

Source: The Roper Reports.

Real vs. Fake

Only 38 percent of American households buy a real Christmas tree to celebrate the holidays.

Source: National Christmas Tree Association.

To Dispose or Not to Dispose

Even though disposable diapers are environmentally harmful, about 85 percent of all parents use them exclusively.

Source: Wertheim, Schroder and Company.

Americans Are Pen Pals

Of the 202 billion cards and letters mailed in the world, about 41 percent were by Americans.

U.S. Postal Service.

Service Performance

In 1988, the U.S. Postal Service's performance was 95 percent on time for local overnight delivery, 86 percent for 2-day delivery within 600 miles, and 89 percent on time for cross-country delivery within 3 days.

Source: Ibid.

Mailmen Get Exercise

The average letter carrier walks an average of 5.2 miles per day on delivery.

Source: National Association of Letter Carriers.

Getting Published

The odds of getting a book published is about 1 in 10,000 with an agent and about 1 in 100,000 without an agent. The average advance is between $2,500 and $5,000.

Source: The National Writers Club.

It's a Small World

Studies suggest that any randomly selected American adult can be linked to any other such adult by an average of only 2 intermediates. If Mr. Smith and Mr. Jones are 2 persons in the United States picked at random, the chances are that Smith will know someone who knows someone who knows Jones. This conclusion is based on the assumption that the "average" American knows about 1,000 people well enough to recognize them on the street and greet them by name.

Source: Ithiel de Sola Pool of the Massachusetts Institute of Technology, and Manfred Kochen of the University of Michigan.

Wheel of Fortune Contestants

While perhaps millions of Americans write in to this popular game show each year to apply to be contestants, an estimated 27,000 are actually interveiwed. Some 585 will get an opportunity to spin the wheel on nighttime airings, and another 676 will do the same on the daytime versions. This means the odds are roughly 27 to 1 against being invited to appear on the show—if you are lucky enough to be asked to audition.

Source: David R. Sams and Robert L. Shook. *Wheel of Fortune*. St. Martin's Press, 1987.

Sales Per Square Foot in U.S. Shopping Malls

There are 348 shopping centers in the United States with 1 million or more square feet. The specialty stores' volume averaged $164 per square foot in 1989.

Source: International Council of Shopping Centers.

Fat Cats

An estimated 20 percent of the nation's cats are overweight. Like humans, obesity increases their chances of heart disease and other fatal maladies.

Source: Heinz Pet Products.

Homebound Earthquakes

Odds of an earthquake in your state in a year (the odds *against* unless otherwise noted):

Alabama	11.0 to 1	Colorado	5.0 to 1
Alaska	4.2 to 1 *for*	Delaware	23.0 to 1
Arizona	3.6 to 1	Georgia	23.0 to 1
Arkansas	11.0 to 1	Hawaii	10.6 to 1 *for*
California	9.2 to 1 *for*	Idaho	2.7 to 1

Illinois	5.0 to 1	New York	5.0 to 1
Maine	11.0 to 1	North Carolina	7.0 to 1
Maryland	11.0 to 1	Ohio	23.0 to 1
Massachusetts	23.0 to 1	Oklahoma	7.0 to 1
Mississippi	11.0 to 1	Oregon	23.0 to 1
Missouri	5.0 to 1	Pennsylvania	11.0 to 1
Montana	*even*	South Carolina	3.8 to 1
Nebraska	23.0 to 1	Tennessee	7.0 to 1
Nevada	3.5 to 1	Texas	3.8 to 1
New Hampshire	23.0 to 1	Utah	*even*
New Jersey	23.0 to 1	Virginia	11.0 to 1
New Mexico	11.0 to 1	Wyoming	1.3 to 1 *for*

Source: U.S. Department of the Interior, Geological Survey, Denver, Colorado.

Homebound Tornadoes

Odds of a tornado on a given day (the odds against):

	Average Number of Tornadoes per Year 1976–1977	Odds of a Tornado Somewhere in the State on Any Given Day
Alabama	25.5	13.3 to 1
Alaska	0	—
Arizona	3.5	103.3 to 1
Arkansas	29.0	11.6 to 1
California	4.5	80.1 to 1
Colorado	37.0	8.9 to 1
Connecticut	0	—
Delaware	2.5	145.0 to 1
Washington, D.C.	0	—
Florida	51.0	6.1 to 1
Georgia	18.5	18.7 to 1
Hawaii	0	—
Idaho	0	—

	Average Number of Tornadoes per Year 1976–1977	Odds of a Tornado Somewhere in the State on Any Given Day
Illinois	30.0	11.2 to 1
Indiana	20.5	16.8 to 1
Iowa	27.0	12.5 to 1
Kansas	15.0	23.3 to 1
Kentucky	8.5	41.9 to 1
Louisiana	28.0	12.0 to 1
Maine	1.0	364.0 to 1
Maryland	3.0	120.7 to 1
Massachusetts	1.5	242.3 to 1
Michigan	32.0	10.4 to 1
Minnesota	18.0	19.3 to 1
Mississippi	34.5	9.6 to 1
Missouri	14.5	24.2 to 1
Montana	4.0	90.2 to 1
Nebraska	47.0	6.8 to 1
Nevada	1.0	364.0 to 1
New Hampshire	1.0	364.0 to 1
New Jersey	1.5	242.3 to 1
New Mexico	5.0	72.8 to 1
New York	6.5	55.1 to 1
North Carolina	25.0	13.6 to 1
North Dakota	41.0	7.9 to 1
Ohio	16.0	21.8 to 1
Oklahoma	41.0	7.9 to 1
Oregon	0	—
Pennsylvania	18.0	19.3 to 1
Puerto Rico	0	—
Rhode Island	0	—
South Carolina	13.0	7.1 to 1
South Dakota	16.0	21.6 to 1

	Average Number of Tornadoes per Year 1976–1977	Odds of a Tornado Somewhere in the State on Any Given Day
Tennessee	12.0	29.4 to 1
Texas	150.5	1.4 to 1
Utah	0	—
Vermont	0	—
Virginia	10.5	33.8 to 1
Virgin Islands	0.5	729.0 to 1
Washington	0	—
West Virginia	1.5	242.3 to 1
Wisconsin	13.5	26.0 to 1
Wyoming	18.5	18.7 to 1

Source: General summary of tornadoes, 1977. U.S. Department of Commerce, National Oceanic and Atmospheric Administration, Environmental Data Service, National Climate Center.

Lefties in the Womb

While about 15 percent of children under age 10 are left-handed, only about 5 percent of fetuses are (they suck their left thumbs).

Source: Queen's University of Belfast, Northern Ireland.

Phobic Drivers

4 million Americans share one of the most debilitating of anxiety disorders: a fear of driving. Once dismissed as a form of hypochondria, anxiety disorders are now considered the most prevalent mental-health problem in the country, possibly affecting as many as 1 out of every 10 people. Fears linked to driving are thought to be among the most common.

Source: *The Wall Street Journal*, October 1990.

Vanity License Plates

About 4 percent, or 1.1 million out of the 25 million cars and trucks in the state of California, have personalized license plates.

Source: California Department of Motor Vehicles.

Driver Donors

21 percent of the drivers in the state of Ohio have given permission on their driver's license to donate organs if they are killed in an automobile accident.

Source: Ohio Bureau of Motor Vehicles, Ohio Department of Health.

Give a Book for Christmas (We Recommend *The Book of Odds*)

During the 1990 holiday season, some 87.5 million Americans gave or received a book. That's 47 percent of all adult Americans. Only 39 percent gave a book during the previous holiday season.

Source: American Booksellers Association.

"Your Check Is in the Mail" from Publishers Clearing House . . .

Do the odds of winning increase when you buy a subscription for one of their magazines? 4 out of 5 people who enter the contest do not buy a magazine subscription and the law is that buyers and nonbuyers have the same chance to win. Of course, you must respond in order for your winning number to win, which means that you must invest in a 29-cent stamp.

Source: Publishers Clearing House.

By the year 2000, 2 out of 3 Americans could be illiterate.

It's true.

Today, 75 million adults...about one American in three, can't read adequately. And by the year 2000, U.S. News & World Report envisions an America with a literacy rate of only 30%.

Before that America comes to be, you can stop it...by joining the fight against illiteracy today.

Call the Coalition for Literacy at toll-free **1-800-228-8813** and volunteer.

Volunteer Against Illiteracy. The only degree you need is a degree of caring.